HUBBLE

3 1571 00282 3568

BY Edward J. Weiler

FOREWORD BY Charles F. Bolden, Jr.

EDITED BY **Robert Jacobs,**

Dwayne Brown, JD Harrington,

Constance Moore, and Bertram Ulrich

Abrams, New York

HUBBLE

A Journey Through Space and Time

CONTENTS

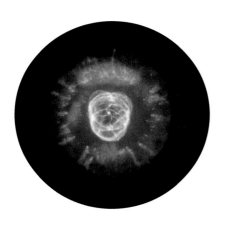

Two thousand and ten marks the twentieth anniversary of one of the most important scientific tools ever employed by humanity, the Hubble Space Telescope. The Hubble telescope is the most celebrated celestial observer since Galileo assembled his first optical instrument four hundred years ago. It was theoretical physicist and astronomer Lyman Spitzer, Jr., who, in 1946, realized that our atmosphere was

an imperfect window on the universe and proposed the idea of placing a large telescope in orbit, free of the atmospheric obstacles that cloud ground-based observatories. He would work for the next five decades on this concept and is credited with being the driving force behind the development of what eventually would become Hubble, launched on April 25, 1990.

Named after the trailblazing astronomer Edwin P. Hubble, the space-based observatory has revolutionized astronomy by providing unprecedented deep and clear views of the universe, ranging from our own solar system to the extremely remote and fledgling galaxies that formed not long after the big bang. As my crewmates were preparing for the STS-31 space shuttle flight that would put Hubble in orbit, there was a lot of discussion about the Big Bang theory, which is credited to the telescope's namesake.

While we were in crew quarters at the Kennedy Space Center in Florida, we talked among ourselves about how significant this mission would be to the future of astrophysics and astronomy. Now, as I look back with the benefit of nearly two decades of hindsight, I realize we grossly underestimated the importance and appeal of this seemingly timeless space observer. Hubble has greatly exceeded anything I imagined as a space explorer. I wish I better understood

how and why it captivates people around the world in a way no other scientific instrument has before. In many ways, I believe it represents our ongoing need to advance the frontiers that expand humanity's reach. Hubble takes us on a journey beyond what we know. It is a time machine that has managed to capture the minds and imaginations of people around the world.

The tragic loss of fourteen astronauts aboard the space shuttles *Challenger* in 1986 and *Columbia* in 2003 underscored the inherent risk in each and every mission. After *Columbia*'s loss threatened a curtailment of the space shuttle program, it appeared Hubble's future might be in jeopardy. I was privileged to serve on the National Research Council committee that looked at saving the telescope, which would cease to function without maintenance. The initial thought was that we would fly a robotic maintenance mission. Teams of scientists and engineers worked earnestly to try to find a way to do that—but it just was not technically feasible at that time. We had to face the limitations of what robotics could do. So the people of NASA stepped up. They are highly intelligent, creative, flexible, and bold in their thinking, and in 2006, the decision was made to move forward with the STS-125 mission to service and repair Hubble one final time with the space shuttle.

In May 2009, just as we had done four times previously, we launched a life-renewing servicing and repair mission to our old friend. The space shuttle *Atlantis* carried seven astronauts and a host of instruments to the ailing observatory. After five dramatic back-to-back spacewalks that included a lot of finesse and a little brute force, a rejuvenated Hubble was released back into orbit, ready to continue work that has fundamentally changed our understanding of the universe.

By design, the Hubble Space Telescope was to be repaired and upgraded with the help of astronauts. It is a powerful example of how humans and machines can work together. Too often the discussion is mistakenly focused on the desire to make a choice between the two. Hubble illustrates how humans and machines can work in harmony for the benefit of exploration.

Beyond its scientific contributions, Hubble's imagery has had a profound impact on the world's imagination and spiritual growth. It is as if Hubble is alive, sharing the dramatic stories of universal creation and destruction by peering back in time, capturing the light that has traveled for billions of years. We owe two decades of astronomical, scientific, and technological breakthroughs to Hubble's history-changing and richly illustrated study of time and space. After a challenging beginning, Hubble is

Messier 74, also called NGC 628, is a stunning example of a "grand design" spiral galaxy that is viewed by Earth observers nearly face-on. Its perfectly symmetrical spiral arms emanate from the central nucleus and are dotted with clusters of young blue stars and glowing pink regions of ionized hydrogen. These regions of star formation show an excess of light at ultraviolet wavelengths. Tracing along the spiral arms are winding dust lanes that also begin very near the galaxy's nucleus and follow along the length of the spiral arms. M74 is located roughly 32 million light-years away, in the direction of the constellation Pisces.

arguably the most successful comeback story and scientific project of all time, in terms of scientific value and its ongoing public appeal. The telescope remains significant by continuing to exploit its unique vantage point where no other observer of the cosmos can compete.

Never before has the universe seemed so gorgeous, powerful, and diverse. Anyone who appreciates the wonders of the universe should take the time to enjoy this beautifully illustrated galactic tour, guided by the scientists, engineers, and astronauts who make Hubble's exciting and ongoing journey of exploration and discovery possible.

Introduction
Edward J. Weiler

Hubble: A Journey Through Space and Time presents some of the greatest scientific discoveries and technological breakthroughs via extraordinary photographs. For me, this book also provides a teachable moment in perseverance and unparalleled teamwork.

Science fiction fascinated me growing up in inner-city Chicago. Wanting to learn more about the stars and universe, I first used a mail-order cardboard telescope to explore the night sky. This crude device could only see a few objects. My second, "real" telescope saw much more, including the moons of Jupiter and the rings of Saturn. About the same time, the U.S. space program was unfolding, and I witnessed science fiction turning into science fact, confirming my desire to become a NASA astronomer.

My place in Hubble's story began after I completed my PhD work at Northwestern University and then went to Princeton to work as a young researcher. Lyman Spitzer, Jr., became my first postgraduate boss. To this day, I often deeply reflect on my personal and professional relationship with the man who shall forever be remembered as the father of Hubble.

Congress officially approved the space telescope project in 1977. Later named in honor of astronomer Edwin P. Hubble, it would be designed with America's space shuttle and servicing in mind. The size of the mirror that collected light to take pictures was a direct result of the size of the shuttle cargo bay. Hubble would be serviced by astronauts and operate at least fifteen years.

I could never have imagined that after being hired by NASA in 1978, I would become Hubble's chief scientist in 1979 and hold that job for twenty years. It was a constant battle for me to keep the project alive. I became a determined force for Hubble supporters and a formidable shield against opponents. Despite the multitude of challenges, I never gave up.

Hubble was originally scheduled to lift off in 1983. Launch was repeatedly delayed due to technical challenges with the telescope. It was finally rescheduled for 1986, when the tragic loss of the space shuttle *Challenger* grounded the shuttle fleet for nearly three years.

Hubble eventually launched aboard the space shuttle *Discovery* in April 1990. Astronomers' dreams were finally fulfilled with a large, optically superb telescope orbiting above the Earth's distorting atmosphere to provide uniquely clear and deep views of the cosmos. While small compared to major ground-based telescopes, Hubble's mirror was the most perfect ever made, with no peaks or valleys greater than half a millionth of an inch. Hubble's optics could see a firefly roughly ten thousand miles away, the distance from Washington, D.C., to Sydney, Australia.

Hubble was equipped with a variety of instruments, including the Wide Field and Planetary Camera, which was the single most important instrument to provide ultrasharp color pictures of the universe. Astronomers eagerly awaited Hubble's first images. For many, this was the culmination of a lifelong dream. However, when the images appeared, they were a nightmare.

During Hubble's initial checkout, scientists discovered it was nearsighted. The supposedly perfect primary mirror had been ground into the wrong shape. This imperfection was one-fiftieth the width of a human hair, but it prevented light from the outer regions of the mirror from coming into focus at the same point as light from the inner regions. In scientific terminology, this is called a spherical aberration. If you had asked me to give you my biggest worries of a possible Hubble problem, this would not have been on my list.

I appeared on numerous news programs telling the world of NASA's failure. Instead of becoming a national treasure, Hubble was a national embarrassment. To me this compared to being on top of Mount Everest then tumbling down to Death Valley. There was one chance to correct this debacle. While the mirror itself could not be fixed, a new camera and other instruments could be developed with corrective optics to cancel out the aberration. In a sense, we would give Hubble glasses.

A second generation Wide Field and Planetary Camera was already under construction. Because of the spherical aberration, it was equipped with special mirrors to counteract the effects of Hubble's flawed main mirror. This camera would eventually save Hubble from being a failure. An instrument composed of small mirrors also was built for the other three instruments on Hubble to counteract the effects of the mirror problem.

It took three and a half years to complete testing and train the astronaut crew for the first and most ambitious servicing mission ever attempted by NASA. The U.S. civilian space program's reputation was on the line. Called "the miracle mission" by many scientists, the space shuttle *Endeavour* launched in December 1993. A record five back-to-back spacewalks were successfully completed, replacing components and correcting Hubble's blurry vision. Hubble was transformed from a comic's joke to the world's greatest space observatory.

To keep Hubble healthy, regular servicing missions were needed to replace failed components, install new hardware, and enhance imaging capabilities. Missions were conducted in 1997, 1999, and 2002. In essence, shuttle astronauts were Hubble's physicians and surgeons.

A final mission to further extend Hubble into the next decade had been planned for 2004. The loss of space shuttle *Columbia* in 2003 changed that. The review board studying the tragedy recommended that future shuttle missions fly in orbits to allow them to reach the International Space Station, a "safe haven" in case of an emergency. A mission to service Hubble could not do this. Because of this and other safety concerns, NASA determined it too dangerous to service Hubble. The mission was canceled.

The decision set off a storm of controversy and criticism among the astronomical community, news media, and key lawmakers. It meant that a more advanced camera and other components, already built, would not be installed. It also left Hubble's continued operation at the mercy of its aging equipment. Options were explored for a robotic servicing mission, but the mission's scope was limited, technical risks high, and projected costs extreme. Without the mission, engineers believed Hubble would not survive past 2010.

NASA resumed shuttle flights in 2005 and, independent of Hubble concerns, devoted considerable resources to developing shuttle repair techniques in space. NASA formed a team to conduct a detailed analysis of performance and procedures necessary to conduct a successful Hubble repair mission. Using results from post-*Columbia* missions, data revealed a safe and effective servicing mission could be accomplished. While there is inherent risk in all spaceflight activities, the desire to preserve an asset like Hubble made a mission prudent. To address the "safe haven" concern, a second orbiter in parallel with the first would be ready for launch for a rescue mission.

In 2006, NASA announced a final servicing mission to Hubble. The new hardware was ready, an astronaut crew selected. The mission was originally planned to launch on October 14, 2008. However, during preparations, a vital Hubble component failed on orbit. Because the hardware was critical, the mission was postponed to allow time to prepare a spare component.

The final servicing mission launched aboard the space shuttle *Atlantis* in May 2009. Astronauts conducted unprecedented repairs and replaced instruments using new, untried techniques. It was bittersweet closure for me to see our astronauts safely return from servicing Hubble one last time. Their success retrofitted Hubble with a full suite of scientific instruments, batteries, gyroscopes, and other equipment, making the telescope more powerful than ever and equipped to explore well into the next decade.

Throughout all of Hubble's challenges, the telescope has continuously rewritten science textbooks. Interestingly, many of those textbooks were the same ones my science colleagues and I used to obtain our PhDs. Hubble has returned a steady stream of spectacular photographs of planets, stars, and galaxies that have found their way into all facets of our society and culture. Pictures abound in school textbooks, art, movies, music, and on the Internet. Thousands of researchers' careers have been shaped by Hubble's images.

Hubble has enabled our minds and spirits to travel billions of light-years to bring the universe up close and personal. A large part of Hubble's success and longevity is due to worldwide public support for the telescope. Hubble is an instantly recognized scientific icon, and its positive impact on the world will continue for decades.

While this book represents a sampling of the telescope's twenty years' worth of contributions, many of Hubble's greatest discoveries are yet to come.

Happy anniversary, Hubble.

The Hubble Space Telescope undergoing final inspection before being released into orbit for the first time by the crew of STS-31 aboard space shuttle *Discovery* in 1990.

Every ninety-seven minutes, the Hubble Space Telescope circles Earth, moving at the speed of about 5 miles per second, which is fast enough to travel across the United States in about ten minutes. As it travels, Hubble's mirror captures light and directs it into its several science instruments.

Hubble is a type of telescope known as a Cassegrain reflector. Light hits the telescope's primary mirror, bounces off, and encounters a secondary mirror. The secondary mirror focuses the light through a hole in the center of the primary mirror that leads to the telescope's science instruments.

People often mistakenly believe that a telescope's power lies in its ability to magnify objects. Telescopes actually work by collecting more light than the human eye can capture on its own. The larger a telescope's mirror, the more light it can collect, and the better its vision. Hubble's primary mirror is 94.5 inches in diameter, small when compared with those of ground-based telescopes, many of which can be 400 inches and larger. However, what makes Hubble unique is its vantage point above the distortion of the atmosphere, giving the observatory remarkable clarity.

Once the mirrors capture the light, Hubble's science instruments work together or individually to provide the observation. Each instrument is designed to examine the universe in a different way. Since May 2009, Hubble's instrumentation is as follows:

The **Wide Field Camera 3 (WFC3)** sees three different kinds of light: near-ultraviolet, visible, and near-infrared, though not simultaneously. Its resolution and field of view are much greater than that of Hubble's other instruments. It can be used to study the effects of dark energy and dark matter on visible light, and the formation of individual stars, and to record the discovery of extremely remote galaxies previously beyond Hubble's vision.

The **Cosmic Origins Spectrograph (COS)** is a spectrograph that sees exclusively in ultraviolet light. Spectrographs act something like prisms, separating light from the cosmos into its component colors. This provides a wavelength "fingerprint" of the object being observed, which tells us about its temperature, chemical composition, density, and motion.

The **Advanced Camera for Surveys (ACS)** sees visible light, and is designed to study some of the earliest activity in the universe. ACS helps map the distribution of dark matter, detects the most distant objects in the universe, searches for massive planets, and studies the evolution of clusters of galaxies.

The **Space Telescope Imaging Spectrograph (STIS)** is a spectrograph that sees ultraviolet, visible, and near-infrared light, and is known for its ability to hunt black holes. While COS works best with small sources of light, such as stars or quasars, STIS can map out larger objects like galaxies.

The **Near Infrared Camera and Multi-Object Spectrometer (NICMOS)** is Hubble's heat sensor. Its sensitivity to infrared light—perceived by humans as heat—lets it observe objects hidden by interstellar dust, like stellar birth sites, and gaze into deepest space.

Finally, the **Fine Guidance Sensors (FGS)** are devices that lock onto "guide stars" and keep Hubble pointed in the right direction. They can be used to precisely measure the distance between stars and their relative motions.

All of Hubble's functions are powered by sunlight. Hubble has solar arrays that convert sunlight directly into electricity. Some of that electricity is stored in batteries that keep the telescope running when it's in Earth's shadow, blocked from the Sun's rays.

HST20

As the space observatory approaches two decades of operation, the editors worked with a group of senior Hubble program scientists to select twenty of the most scientifically and culturally important Hubble observations and discoveries to date. The HST20 appear throughout this book on gray panels like this one.

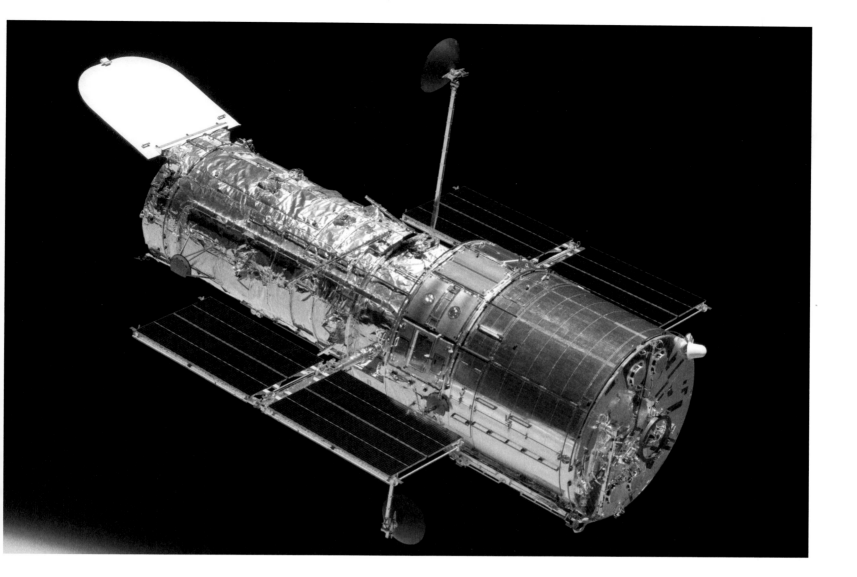

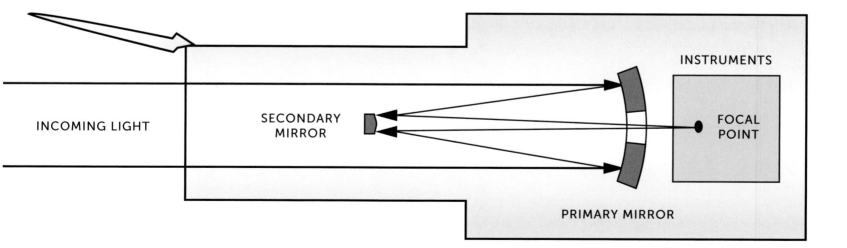

INCOMING LIGHT

SECONDARY
MIRROR

INSTRUMENTS

FOCAL
POINT

PRIMARY MIRROR

TOP: The Hubble Space Telescope was photographed by a crewmember of STS-125 aboard the space shuttle *Atlantis* as the two spacecraft separated on May 19, 2009, after having been linked together for the better part of a week.

ABOVE: When light strikes the concave primary mirror of the Hubble Space Telescope, it is reflected to the convex secondary mirror, then back through a hole in the center of the primary mirror. There, the light comes to the focal point and passes to one of Hubble's instruments. Telescopes of this design are called Cassegrain telescopes, after the person who designed the first one.

In 2009, Hubble trained its new camera, the Wide Field Camera 3, on Jupiter, and took the sharpest visible-light picture of the planet since the *New Horizons* spacecraft flew by it in 2007. Each pixel in this high-resolution image spans about 74 miles (119 kilometers) in Jupiter's atmosphere. Jupiter was more than 370 million miles (600 million kilometers) from Earth when the images were taken. The dark smudge at bottom right is debris from a comet or asteroid that plunged into Jupiter's atmosphere and disintegrated.

Next to sending dedicated spacecraft to visit the various planets in our solar system, the Hubble Space Telescope's high-resolution images may be the best way to study our closest celestial neighbors. Periodic and ongoing monitoring of the planets' atmospheres and surfaces show scientists how these bodies evolve and how they react to unexpected events.

In July 1994, remnants of Comet P/Shoemaker-Levy 9 collided with Jupiter. This was the first collision of two solar system bodies ever to be observed, and the effects of the comet's impact on the Jovian surface captured by Hubble exceeded everyone's expectations. The collisions left giant, dark, scarlike blemishes on Jupiter for weeks. Studies with Hubble have also allowed scientists to discover a number of unexpected components of Jupiter's atmosphere, such as hydrogen sulfide.

Following the successful space shuttle servicing mission of STS-125 in May 2009, Hubble once again captured a violent collision on Jupiter. During the checkout and calibration of the newly restored observatory in July 2009, Hubble captured stunning photographs after an object plunged into the Jovian atmosphere and disintegrated.

Whether studying collisions on Jupiter, discovering new moons orbiting Pluto, capturing the first visible-light images of rings around the planet Uranus, or monitoring dust storms and weather patterns on Mars, Hubble's unique abilities to investigate and monitor these and other events on nearby planets and their satellites can reveal much about our solar system. Hubble has beautifully confirmed the dynamic nature of planetary atmospheres and allowed scientists to more accurately follow their behavior as they study what drives climates on other bodies in our solar system.

In this rare Hubble image of a feature on the Moon, a crisp bird's-eye view clearly shows the ray pattern of bright dust ejected out of the crater Copernicus more than one billion years ago. Hubble can resolve features as small as 600 feet across in the terraced walls of the crater and the hummocklike blanket of material blasted out by the meteor impact.

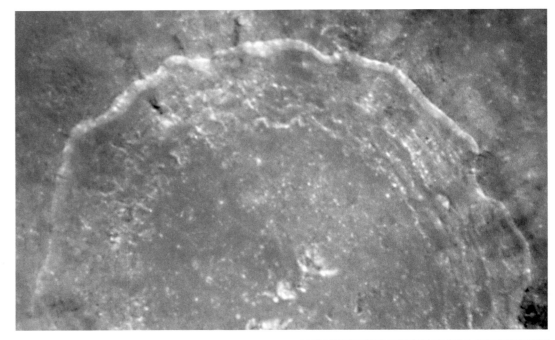

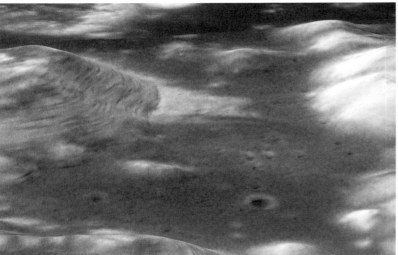

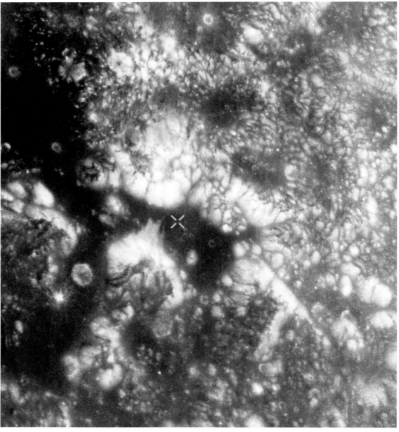

Hubble's exquisite resolution and sensitivity to ultraviolet light, which is reflected off the Moon's surface materials, have allowed scientists to use the observatory to begin to search for the presence of minerals that may be critical for the establishment of sustained human presence on the Moon. The high-resolution ultraviolet and visible-light image above shows the *Apollo 17* landing region within the Taurus-Littrow valley of the Moon in 2005. Humans last walked and drove on the lunar surface in this region (marked X in the image) in December 1972. The Hubble Space Telescope Lunar Exploration team constructed a perspective view looking from west to east up the valley (above left) by overlaying the Hubble image with a digital-terrain model acquired by the Apollo program. These images help the team to discriminate lunar materials enriched in ilmenite, a titanium-bearing oxide of potential value as a resource in human exploration of the Moon.

An ultraviolet-light image of Venus, taken when the planet was 70.6 million miles (113.6 million kilometers) from Earth, captures details in its cloud cover. Venus is covered with clouds made of sulfuric acid, rather than the water-vapor clouds found on Earth. These clouds permanently shroud Venus's volcanic surface, which has been radar mapped by spacecraft and from Earth-based telescopes. At ultraviolet wavelengths cloud patterns become distinctive. In particular, a horizontal Y-shaped cloud feature is visible near the equator. Similar features were seen from the *Mariner 10*, *Pioneer Venus*, and *Galileo* spacecrafts. This global feature might indicate atmospheric waves, analogous to high and low pressure cells on Earth.

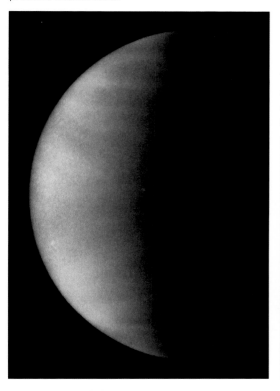

This panoramic composite of Hubble images made during the close approach of Mars to the Earth in December 2007 spans 360 degrees of Mars's surface, starting at a longitude line of 230 degrees on the left edge. It spans nearly to the north and south polar cap regions in Mars's latitude.

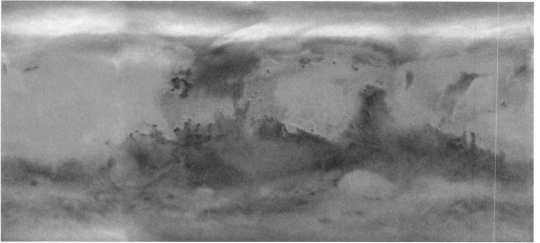

Taken at the beginning of a long-term program to monitor seasonal and interannual changes occurring on the surface and in the atmosphere of Mars in 1991, this is among the sharpest images of Mars ever captured from Earth. It reveals Martian atmospheric features, as well as surface and topographic details ranging from large impact basins down to surface markings as small as 31 miles (50 kilometers) across.

Hubble participated in NASA's Deep Impact mission, which sent a spacecraft to impact Comet 9P/Tempel 1 on July 4, 2005, to probe its interior. This image, taken by Hubble on the morning of June 30, shows an undisturbed and quiet comet. The dusty inner coma around the nucleus of the comet is visible, but the solid nucleus itself is below Hubble's resolution. The nucleus was at a distance of 83 million miles (134 million kilometers) when the image was taken. As the telescope was locked on the movement of the comet, the background stars left small trailed arcs during the time the exposures were taken.

As Comet P/Shoemaker-Levy 9 hurtled toward a July 1994 collision with the giant planet Jupiter, Hubble resolved it into approximately twenty objects resembling a string of pearls. Hubble showed that the comet's nuclei were much smaller than originally estimated from observations with ground-based telescopes. The Hubble observations showed that the nuclei were probably less than 3 miles (5 kilometers) across, as opposed to earlier estimates of 9 miles (14 kilometers).

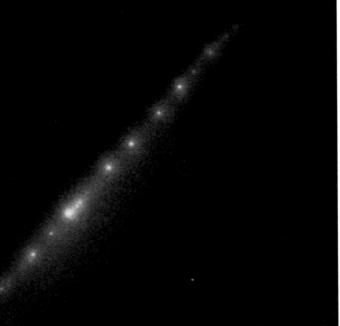

SHOEMAKER-LEVY 9 JUPITER IMPACT

Hubble provided a ringside seat to what was previously thought to be a once-in-ten-thousand-year event when two dozen chunks of Comet P/Shoemaker-Levy 9 (opposite bottom) smashed into Jupiter in July 1994. The telescope snapped dramatic images of massive explosions that accompanied each impact, below. The impacts sent towering mushroom-shaped fireballs of hot gas into the Jovian sky. The doomed comet had been pulled apart two years earlier by Jupiter's gravity. Each impact left temporary black, sooty scars in Jupiter's planetary clouds. Astronomers were surprised in 2009 when a single comet fragment hit Jupiter and left a similar-looking scar (see page 16). Apparently single comet impacts happen more frequently. Credit: H. Hammel (MIT)

In this crisp Hubble image, Jupiter's moon Ganymede is shown just before it ducks behind the giant planet. Composed of rock and ice, Ganymede is the largest moon in our solar system. It is even larger than the planet Mercury. But Ganymede looks like a dirty snowball next to Jupiter, the largest planet. Jupiter is so big that only part of its southern hemisphere can be seen in this image. Hubble's view is so sharp that astronomers can see features on Ganymede's surface, most notably the white impact crater, Tros, and its system of rays, bright streaks of material blasted from the crater. Tros and its ray system are roughly the width of Arizona. The image also shows Jupiter's Great Red Spot, the large eye-shaped feature at center left. A storm the size of two Earths, the Great Red Spot has been raging for more than three hundred years.

In the spring of 2008, a third red spot appeared alongside its cousins—the Great Red Spot and Red Spot, Jr.—in the turbulent Jovian atmosphere. This third red spot, which is a fraction of the size of the two other features, lies to the west of the Great Red Spot in the same latitude band of clouds. The new red spot was previously a white oval-shaped storm. The change to a red color indicates its swirling storm clouds are rising to heights similar to the clouds of the Great Red Spot. One possible explanation is that the red storm is so powerful it dredges material from deep beneath Jupiter's cloud tops and lifts it to higher altitudes where solar ultraviolet radiation— via some unknown chemical reaction—produces the familiar brick color.

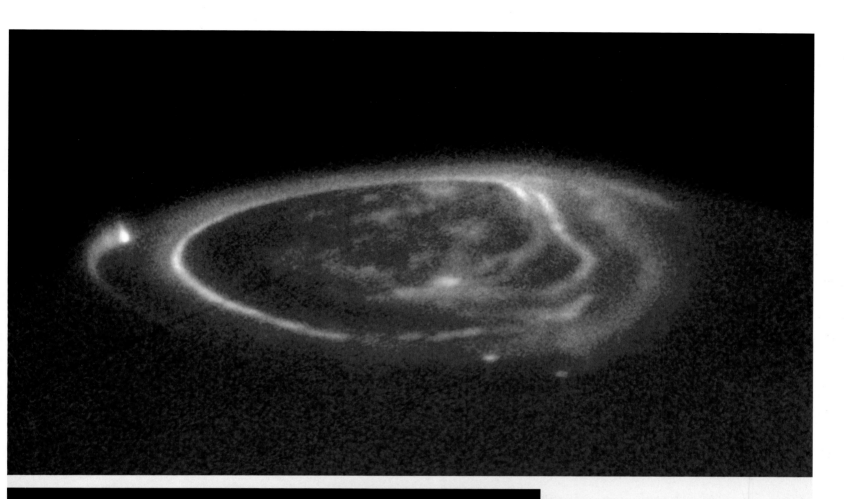

AURORA ON JUPITER

Hubble captured a curtain of glowing gas wrapping around Jupiter's north and south poles like a lasso. This aurora is produced when high-energy electrons race along the planet's magnetic field and into the upper atmosphere, where they excite atmospheric gases, causing them to glow. The Hubble image above, taken in ultraviolet light, also shows the glowing magnetic "footprints" of three of Jupiter's largest moons: Io, Ganymede, and Europa.

Credit: John Clarke (University of Michigan)

October 1998

OPPOSITE: Looming like a giant flying saucer in our outer solar system, Saturn puts on a show as the planet and its magnificent ring system nod majestically over the course of its twenty-nine-year journey around the Sun. The Hubble images below, captured from 1996 to 2000, show Saturn's rings opening up from just past edge on to nearly fully open as it moves from autumn toward winter in its northern hemisphere. Saturn's equator is tilted relative to its orbit by 27 degrees, very similar to the 23-degree tilt of the Earth. As Saturn moves along its orbit, first one hemisphere, then the other is tilted toward the Sun. This cyclical change causes seasons on Saturn, just as the changing orientation of Earth's tilt causes seasons on our planet. Above, the central image of the sequence is enlarged to show more detail.

RIGHT: Saturn's comparatively paper-thin rings are tilted edge on to Earth every fifteen years. Because the orbits of Saturn's major satellites are in the ring plane, too, this alignment gives astronomers a rare opportunity to capture a truly spectacular parade of celestial bodies crossing the face of Saturn. Leading the parade is Saturn's giant moon Titan—larger than the planet Mercury. The frigid moon's thick nitrogen atmosphere is tinted orange with the smoggy by-products of sunlight interacting with methane and nitrogen. Several of the much smaller icy moons that are closer in to the planet line up along the upper edge of the rings. Hubble's exquisite sharpness also reveals Saturn's banded cloud structure.

This August 2003 view of Uranus reveals the planet's faint rings and several of its satellites. The area outside Uranus was enhanced in brightness to reveal the faint rings and satellites. The outermost ring, made of dust and small pebbles, is brighter on the lower side, where it is wider. The bright satellite in the lower right corner is Ariel, which has a snowy white surface. Five small satellites with dark surfaces can be seen just outside the rings. Clockwise from the top, they are: Desdemona, Belinda, Portia, Cressida, and Puck.

In August 2007 Hubble captured a rare view of the entire ring system of the planet Uranus, tilted edge on to Earth. The edge-on rings appear as spikes above and below the planet. Earthbound astronomers only see the rings' edge every forty-two years, as the planet follows a leisurely eighty-four-year orbit about the Sun. However, the last time the rings were tilted edge on to Earth, astronomers didn't even know they existed.

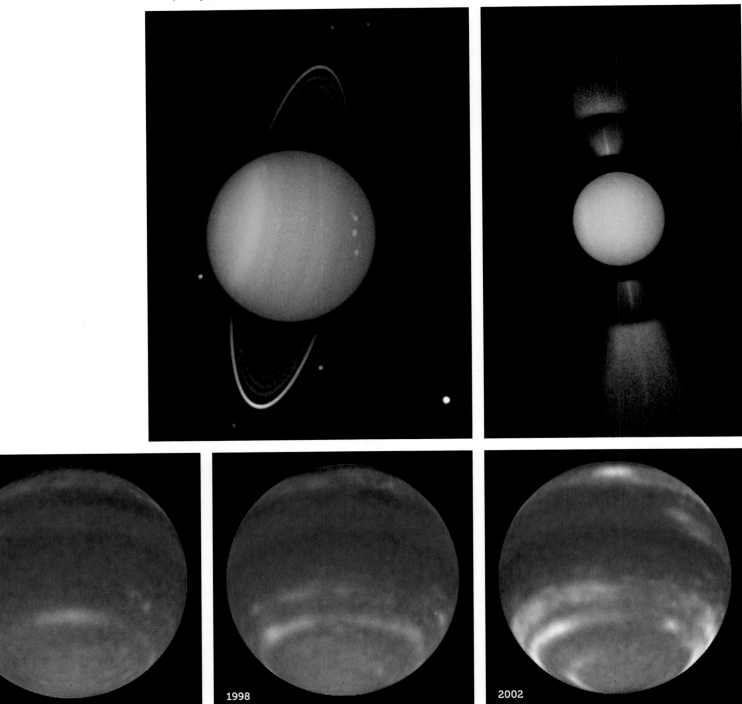

1996

1998

2002

Springtime is blooming on Neptune! This might sound absurd, because Neptune is the farthest and coldest of the major planets, but observations by Hubble made over six years show a distinct increase in the amount and brightness of the clouds encircling the planet's southern hemisphere. Astronomers consider this increase a harbinger of seasonal change.

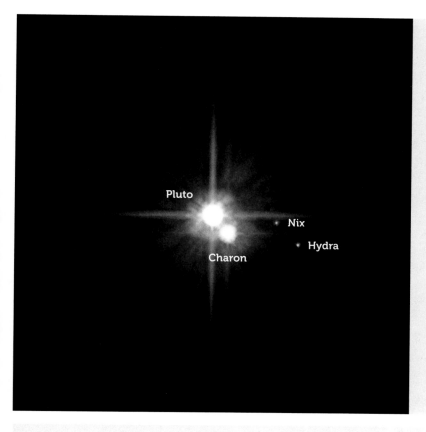

MOONS OF PLUTO

Hubble discovered two new moons orbiting the icy dwarf planet Pluto. Named Nix and Hydra, the moons have the same color as Charon, Pluto's only other known moon. The moons' common color further reinforces the idea that all three moons were born from a single titanic collision between Pluto and another similarly sized Kuiper Belt object billions of years ago.

Credit: H. Weaver (JHU/APL), A. Stern, and the HST Pluto Companion Search Team

MEASUREMENT OF KUIPER BELT OBJECTS

Hubble has searched the solar system's last frontier, a vast outer region called the Kuiper Belt, for relics from the birth of the planets. An icy Kuiper Belt object (KBO), dubbed Quaoar by its discoverers in 2002, is the largest body found in the solar system since the discovery of Pluto seventy-two years ago. Below left is Hubble's photograph of Quaoar, and below right is an artist's vision of how it might look close up. Quaoar is about half the size of Pluto. In 2006, Hubble determined that the diameter of Eris, another KBO discovered in 2005, is slightly larger than Pluto. This discovery eventually helped lead to a spirited debate among astronomers that demoted Pluto to a "dwarf planet." Credit: Mike Brown (Caltech)

Ever since the dawn of humanity we have gazed into the night sky and pondered the meaning of the thousands of stars twinkling overhead. Even casual stargazers, using the most basic optical viewers, recognize that stars are colorful and diverse. Stars come in many different sizes, from about one-tenth the mass of our Sun to more than one hundred times its mass, and colors, from deep red to bluish white.

A star is best described as a ball of gas, usually hydrogen and helium, with enough mass (a minimum of roughly 1.5×10^{29} kg or 27,000 times the mass of the Earth) to sustain nuclear fusion at its core. It is held together by its own gravity. That force continually works to collapse the gas of the star, but it is countered by outward pressures resulting from the nuclear furnace at the star's center, and all stars eventually burn out, sometimes with spectacular results. Hubble observes stars at various periods of their lives—birth, middle age, and death—and has enabled a much deeper understanding of stars' life cycles and the circumstances that lead to their creation and destruction.

This celestial life cycle drives cosmic evolution and the recycling of elements over billions of years. As Hubble has demonstrated with its stellar and interstellar studies that serve the art of photography as well as science, it is a process that is both violent and beautiful to witness. In what has become one of the most famous images of modern times and an inspiration to millions of viewers, the Hubble Space Telescope in 1995 captured what is known as the "Pillars of Creation," a star-forming region in the Eagle Nebula about 6,500 light-years from Earth.

Thousands of sparkling young stars are nestled within the giant nebula NGC 3603, a stellar "jewel box" that is one of the most massive young star clusters in the Milky Way galaxy. NGC 3603 is a prominent star-forming region in the galaxy's Carina spiral arm, about 20,000 light-years away. Sir John Herschel first discovered the nebula in 1834. This image spans roughly 17 light-years.

BIRTH OF STARS IN THE EAGLE NEBULA

Eerie, dramatic pictures from the Hubble show newborn stars emerging from dense, compact pockets of interstellar gas called evaporating gaseous globules (EGGs) that lie in the Eagle Nebula, a nearby star-forming region 7,000 light-years from Earth in the constellation Serpens. The columns—dubbed "elephant trunks"—protrude from the wall of a vast cloud of molecular hydrogen like stalagmites rising above the floor of a cavern. Inside the gaseous towers, which are light-years long, the interstellar gas is dense enough to collapse under its own weight, forming young stars that continue to grow as they accumulate more and more mass from their surroundings. Credit: Jeff Hester (Arizona State University)

Resembling a nightmarish beast rearing its head from a crimson sea, this monstrous object is actually an innocuous pillar of gas and dust, and an incubator for developing stars. Called the Cone Nebula (NGC 2264)—so named because, in ground-based images, it has a conical shape—this giant pillar resides in a turbulent star-forming region. This picture shows the upper 2.5 light-years of the nebula, a height that equals 23 million round-trips to the Moon. The entire nebula is 7 light-years long and resides 2,500 light-years away, in the constellation Monoceros.

Hubble astronomers uncovered a population of infant stars in the Small Magellanic Cloud, the Milky Way's satellite galaxy 200,000 light-years away. Although star birth is common within the disk of our galaxy, this smaller companion galaxy is more primeval in that it lacks a large percentage of the heavier elements that are forged in successive generations of stars through nuclear fusion. Yet Hubble's exquisite vision detected a population of infant stars embedded in the nebula NGC 346 that are still forming from gravitationally collapsing gas clouds. They have not yet ignited their hydrogen fuel to sustain nuclear fusion. The smallest of these young stars is only half the mass of our Sun.

ABOVE: Star cluster NGC 602 is another star-forming region in the Small Magellanic Cloud. Bright, blue, newly formed stars are blowing a cavity in this nebula, sculpting the inner edge of its outer portions, slowly eroding it away and eating into the material beyond. The diffuse outer reaches of the nebula prevent the energetic outflows from streaming away from the cluster. Elephant trunk–like dust pillars point toward the hot blue stars and are telltale signs of their eroding effect. Star formation started at the center of the cluster and propagated outward, with the youngest stars still forming today along the dust ridges.

OVERLEAF: The core of the spectacular globular cluster Omega Centauri glitters with the combined light of 2 million stars. The entire cluster contains 10 million stars, and is among the biggest and most massive of some two hundred globular clusters orbiting the Milky Way galaxy. Globular clusters are ancient swarms of stars united by gravity. Astronomers at the Max-Planck-Institute for Extraterrestrial Physics in Germany and the University of Texas at Austin have reported on the possible detection of an intermediate-mass black hole in the core of Omega Centauri. Hubble images like this one help pinpoint the center of the cluster, as well as measuring the amount of starlight at the cluster center.

A colorful assortment of stars can be seen in this close-up view of the crowded core of the massive globular cluster Omega Centauri. The stars in Omega Centauri are between 10 billion and 12 billion years old and lie about 16,000 light-years from Earth.

Like a whirl of shiny flakes sparkling in a snow globe, many hundreds of thousands of stars move about in M13, one of the brightest and best-known globular clusters in the northern sky. Although located more than 25,000 light-years away, this glittering metropolis of stars is easily found in the winter sky, in the constellation Hercules.

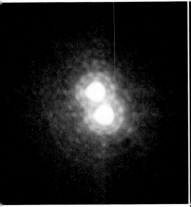

The small open star cluster Pismis 24 lies in the core of the large emission nebula NGC 6357 in Scorpius, about 8,000 light-years away from Earth. Some of the stars in this cluster are extremely massive and emit intense ultraviolet radiation. The brightest object in the picture is designated Pismis 24-1. It was once thought to weigh as much as 200 to 300 solar masses. This would not only have made it by far the most massive known star in the galaxy, but would have put it considerably above the currently believed upper mass limit of about 150 solar masses for individual stars. However, high-resolution Hubble images of the star, above, taken in April 2006 show that it is really two stars orbiting each other. They are each estimated to be 100 solar masses.

In January 2002, a dull star in an obscure constellation suddenly became 600,000 times more luminous than our Sun, temporarily making it the brightest star in our Milky Way galaxy. Light from the star, called V838 Monocerotis, propagated outward through a cloud of dust. In a phenomenon called a light echo, the light reflected or "echoed" off the dust and then traveled to Earth. Because of the extra distance the scattered light travels, it reaches the Earth long after the light from the stellar outburst itself. Therefore, a light echo is an analog of a sound echo produced, for example, when sound from an Alpine yodeler echoes off of the surrounding mountain-sides. Detailed images like these provided astronomers with a CAT scan–like probe of the three-dimensional structure of the shells of dust that surrounded the aging star. The mysterious star has long since faded back into obscurity.

April 30, 2002

October 2004

December 17, 2002

September 2006

A planet bonanza was uncovered during a Hubble survey called the Sagittarius Window Eclipsing Extrasolar Planet Search (SWEEPS). This image is a color composite of images that shows sixteen host stars with extrasolar planet candidates. Hubble peered at 180,000 stars in the crowded central bulge of our galaxy 26,000 light-years away.

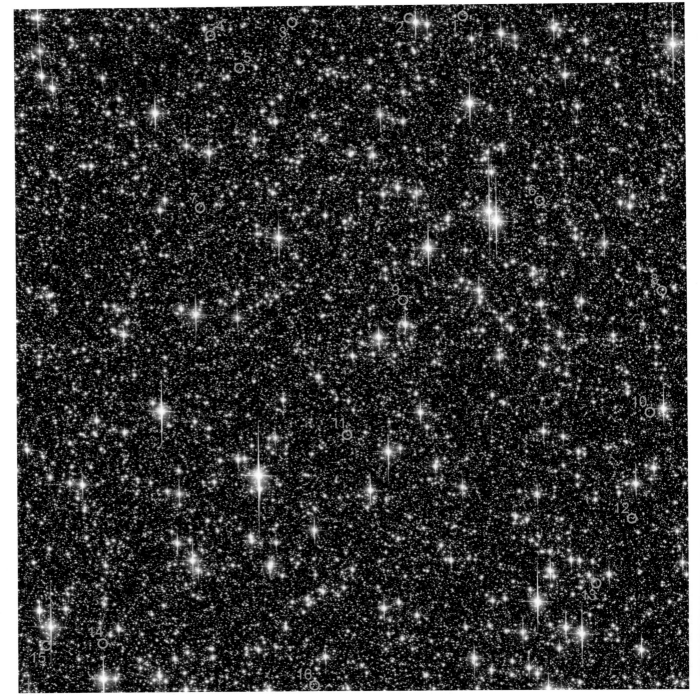

BIRTH OF PLANETS

Hubble observations confirmed that planets form in dust disks around stars. The telescope also witnessed the early stages of planet formation when it observed a blizzard of particles around a star. A survey of the Orion Nebula found 153 disks surrounding newborn stars. The existence of so many young stars with protoplanetary disks mathematically increases the likelihood of other planetary systems. Hubble astronomers estimate the masses of the disks as at least 0.1 to 730 times the mass of our Earth. Credit: NASA, C. R. O'Dell and S. K. Wong (Rice University)

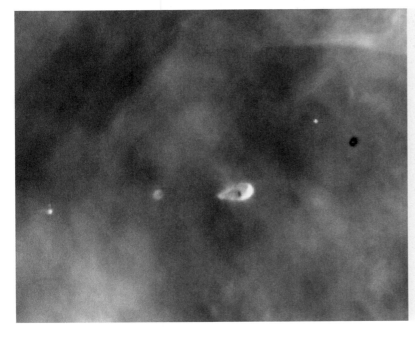

IMAGE OF AN EXOPLANET

Hubble took the first visible-light snapshot of a planet orbiting another star. The images show the planet, named Fomalhaut b, as a tiny point source of light orbiting the nearby bright southern star Fomalhaut, located 25 light-years away in the constellation Piscis Australis. An immense debris disk about 21.5 billion miles (34.6 billion kilometers) across surrounds the star. Fomalhaut b is orbiting 1.8 billion miles (2.9 billion kilometers) inside the disk's sharp inner edge. Estimated to be three times the mass of Jupiter, the planet is brighter than expected. One possibility is that it has a huge Saturn-like ring of ice and dust reflecting starlight. In this image, light from the star, indicated by a white dot, has been blocked so that the faint signal from the planet can be seen. Credit: P. Kalas (UC Berkeley)

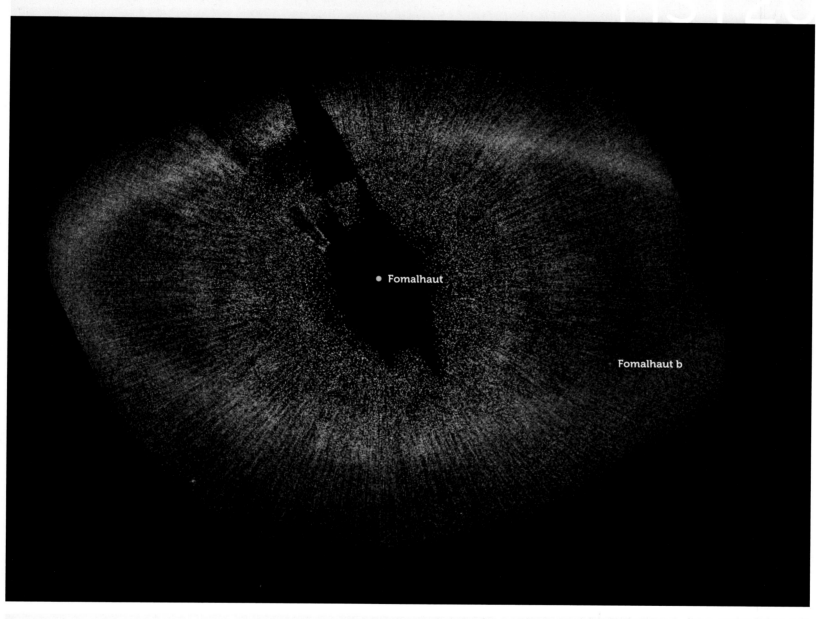

Fomalhaut

Fomalhaut b

COMPOSITION OF EXOPLANET ATMOSPHERE

In 2008 Hubble made the first detection ever of an organic molecule in the atmosphere of a planet orbiting another star. This is an important step in eventually identifying signs of life on a planet outside our solar system. The molecule found by Hubble is methane, which under the right circumstances can indicate the presence of life. In 2001, Hubble made the first detection of the chemical composition of an exoplanet atmosphere by identifying sodium. Hubble demonstrated that it is possible to measure the chemical makeup of extrasolar planet atmospheres and to search for chemical markers of life beyond Earth. Credit: D. Charboneau (Center for Astrophysics), M. Swain (JPL)

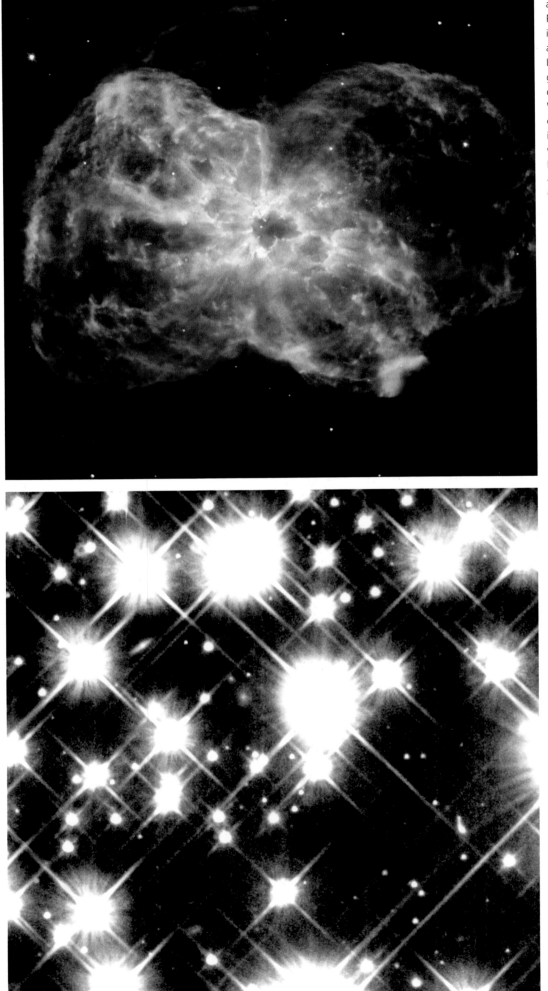

This Hubble image shows the colorful "last hurrah" of a star like our Sun 4,000 light-years away in the direction of the constellation Puppis. The star is ending its life by casting off its outer layers of gas, which form a cocoon around the star's remaining core. Ultraviolet light from the dying star makes the material glow. The burned-out star, called a white dwarf, is the white dot in the center. Our Milky Way galaxy is littered with these stellar relics, called planetary nebulae. The planetary nebula in this image is called NGC 2440, and the white dwarf at its center is one of the hottest known, with a surface temperature of nearly 400,000 degrees Fahrenheit (222,200 degrees Celsius).

Pushing the limits of its powerful vision, NASA's Hubble Space Telescope uncovered the oldest burned-out stars in our Milky Way galaxy. The faintest objects in the image at left are extremely old, dim stars provide a completely independent reading of the universe's age without relying on measurements of the universe's expansion. The ancient white dwarf stars turn out to be 12 to 13 billion years old. Because earlier Hubble observations show that the first stars formed less than one billion years after the universe's birth in the Big Bang, finding the oldest stars put astronomers well within reach of calculating the absolute age of the universe.

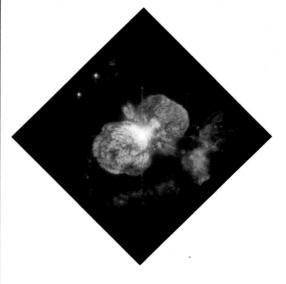

A huge, billowing pair of gas and dust clouds is captured in this stunning image of the supermassive star Eta Carinae, left. Even though Eta Carinae is more than 8,000 light-years away, structures only 10 billion miles across (about the diameter of our solar system) can be distinguished. Dust lanes, tiny condensations, and strange radial streaks all appear with unprecedented clarity.

The large image of Eta Carina was made by Hubble in 1996; the new image from 2009, below, resulted from an investigation that revealed that the amount of Eta Carina's material being carried away by stellar wind is the equivalent of one Sun every thousand years.

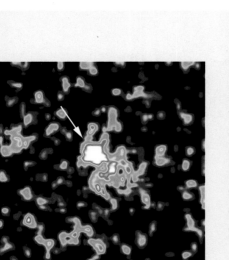

DISCOVERY OF THE ORIGIN OF GAMMA-RAY BURSTS

Hubble found that the universe's most powerful explosions since the Big Bang, gamma-ray bursts, occur among the normal stellar population inside galaxies, and therefore are cosmological in origin. Hubble images of the locations of these daily events showed that these brief flashes of radiation come from far-flung galaxies that are forming stars at enormously high rates. Hubble's observations confirmed that the bursts of light originated from the collapse of massive stars. Astronomers using Hubble also found that a certain type of extremely energetic gamma-ray burst is more likely to occur in galaxies with fewer heavy elements such as carbon and oxygen. Credit: K. Sahu, M. Livio, L. Petro, D. Macchetto (STScI)

SUPERNOVA 1987A

Even though Hubble wasn't launched until 1990, its observations of the remains of a star that exploded in 1987 provided important new insights into the death of a massive star, below right. A supernova is the explosive demise of a massive star that collapses under the weight of its own gravity. The collapsed star then blows its outer layers into space in an explosion that can briefly outshine its entire parent galaxy. Observations of Supernova 1987A, which exploded on February 23, 1987, showed that the early phases following a star's detonation are more complex than was earlier thought. Among Hubble's findings are three mysterious rings of material encircling the doomed star, left. The telescope also resolved brightened spots on the middle ring's inner region, caused by an expanding wave of material from the explosion slamming into it, below left. Credit: R. Kirshner (Center for Astrophysics)

THE CRAB NEBULA

The Crab Nebula is a 6-light-year-wide expanding remnant of a star's supernova explosion. Japanese and Chinese astronomers recorded this violent event nearly one thousand years ago in 1054, as did, almost certainly, Native Americans. The orange filaments are the tattered remains of the star and consist mostly of hydrogen. The neutron star embedded in the center of the nebula, which has a mass equivalent to the Sun crammed into a rapidly spinning ball of neutrons 12 miles (19.3 kilometers) across, is the dynamo powering the nebula's eerie interior bluish glow. The blue light comes from electrons whirling at nearly the speed of light around magnetic field lines from the neutron star. Credit: Jeff Hester, Arizona State University

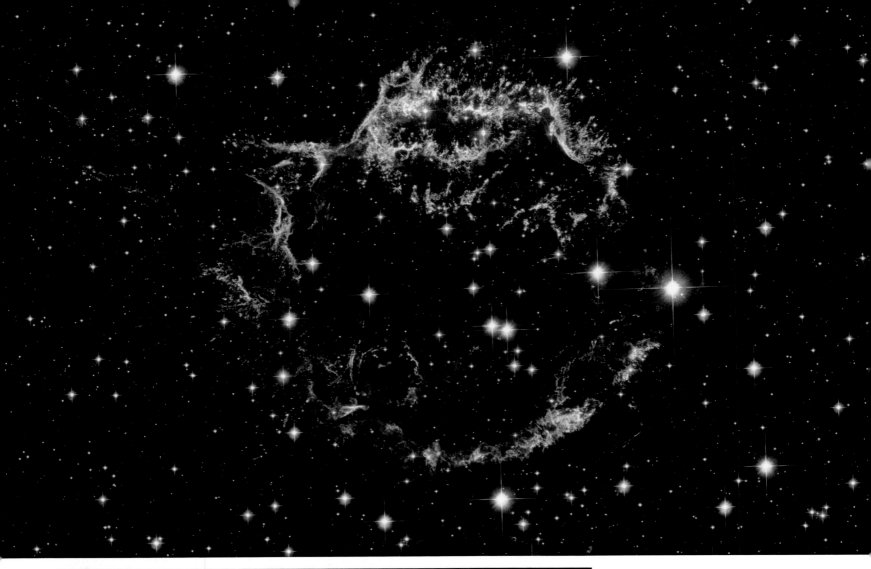

ABOVE: The tattered remains of a supernova explosion known as Cassiopeia A are the youngest known from a supernova explosion in the Milky Way. Cassiopeia A is estimated to be only about 340 years old, located 10,000 light-years away from Earth in the constellation of Cassiopeia. Supernova explosions are the main source of elements more complex than oxygen, which are forged in the extreme conditions produced in these events.

LEFT: A delicate ribbon of gas floats eerily in our galaxy. Could it be a contrail from an alien spaceship, or possibly a jet from a black hole? Actually this Hubble image is a very thin section of a supernova remnant caused by a stellar explosion that occurred more than one thousand years ago. Today we know that SN 1006 has a diameter of nearly 60 light-years, and it is still expanding at roughly 6 million miles per hour. Even at this tremendous speed, it still takes observations separated by years to see significant outward motion of the shock wave against the grid of background stars. In this Hubble image, the supernova would have occurred far off the lower right corner of the image, and the motion would be toward the upper left.

Resembling the puffs of smoke and sparks from a summer fireworks display, these delicate filaments are actually a supernova remnant within the Large Magellanic Cloud, a nearby small companion galaxy to the Milky Way visible from the southern hemisphere. Denoted N 49, or DEM L 190, this remnant is from a massive star that died in a supernova blast whose light would have reached Earth thousands of years ago. This filamentary material will eventually be recycled into building new generations of stars. Our own Sun and planets are constructed from similar debris of supernovae that exploded in the Milky Way galaxy billions of years ago.

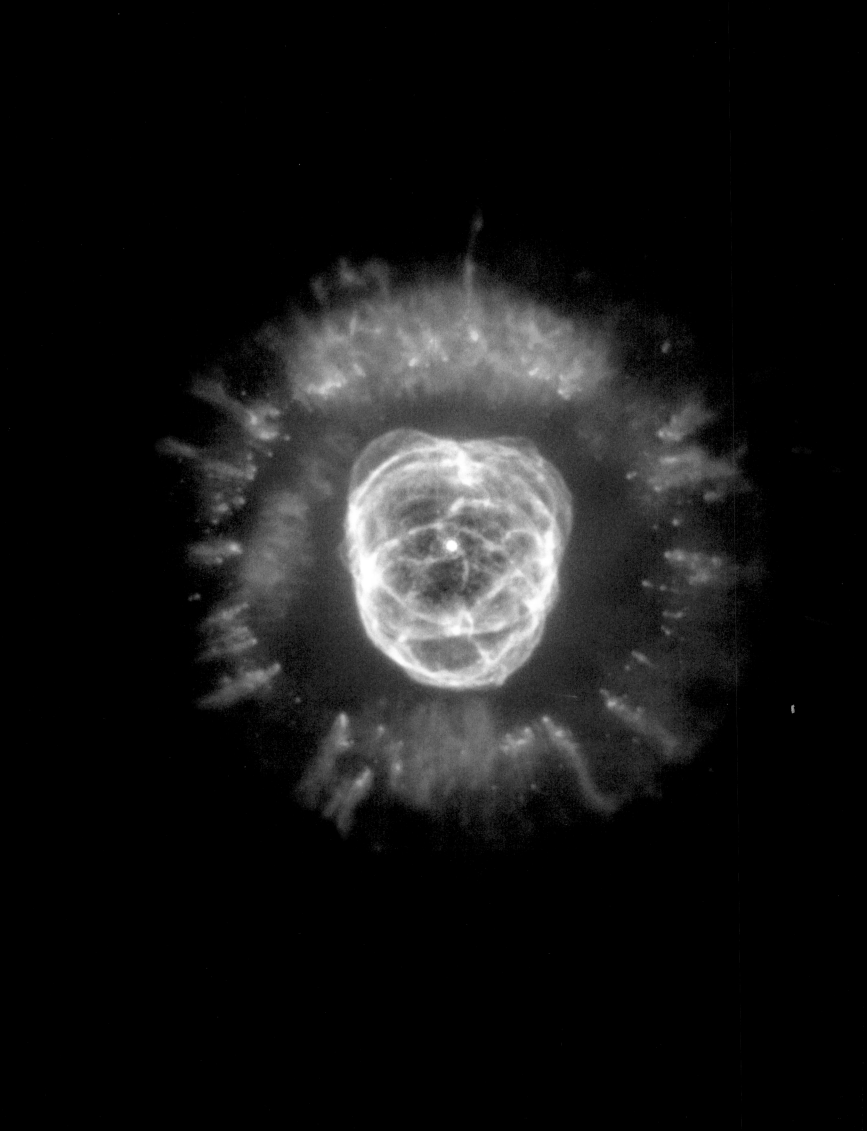

A nebula is a cosmic cloud of gas and dust. Such clouds can contain truly staggering amounts of material. For example, the Orion Nebula has a mass some 2,000 times that of the Sun (or roughly 4×10^{33} kg). It was not until the development of larger telescopes and the introduction of astronomical spectroscopy—the ability to split electromagnetic radiation

including visible light, into its constituent parts, which helps determine the chemical composition of the object emitting the radiation—in the late nineteenth century that astronomers could distinguish between true interstellar dust and gas clouds and compact, unresolved clusters of stars that made up what appeared to be nebulae but were really galaxies.

These celestial clouds contain the elements from which new stars and solar systems are formed. Nebulae consist of vast clouds of gas ionized by the radiation of massive stars or by the explosions of supernovae or by the death throes of ancient stars, which expel their outer atmospheres and then illuminate them. They are to be counted among the largest objects in the galaxy, and they are among the most beautiful objects in the cosmos, often appearing as mist with spirals, loops, wisps, and other intricate shapes.

Hubble's study of nebulae, the material and gasses that are both the products of stellar death and also the building blocks of new stars, explores the universe's continual intertwined processes of destruction and creation.

This stellar relic, first spied by William Herschel in 1787, is nicknamed the "Eskimo" Nebula (NGC 2392) because, when viewed through ground-based telescopes, it resembles a face surrounded by a fur parka. In this Hubble telescope image, the "parka" is really a disk of material embellished with a ring of comet-shaped objects, with their tails streaming away from the central, dying star. Although the Eskimo's "face" resembles a ball of twine, it is, in reality, a bubble of material being blown into space by the central star's intense "wind" of high-speed material. This object is an example of a planetary nebula, so named because many of them have a round appearance resembling that of a planet when viewed through a small telescope. A planetary nebula forms when dying Sun-like stars eject their outer gaseous layers, which then become bright nebulae with amazing and confounding shapes. The Eskimo Nebula is about 5,000 light-years from Earth in the constellation Gemini and began forming about 10,000 years ago.

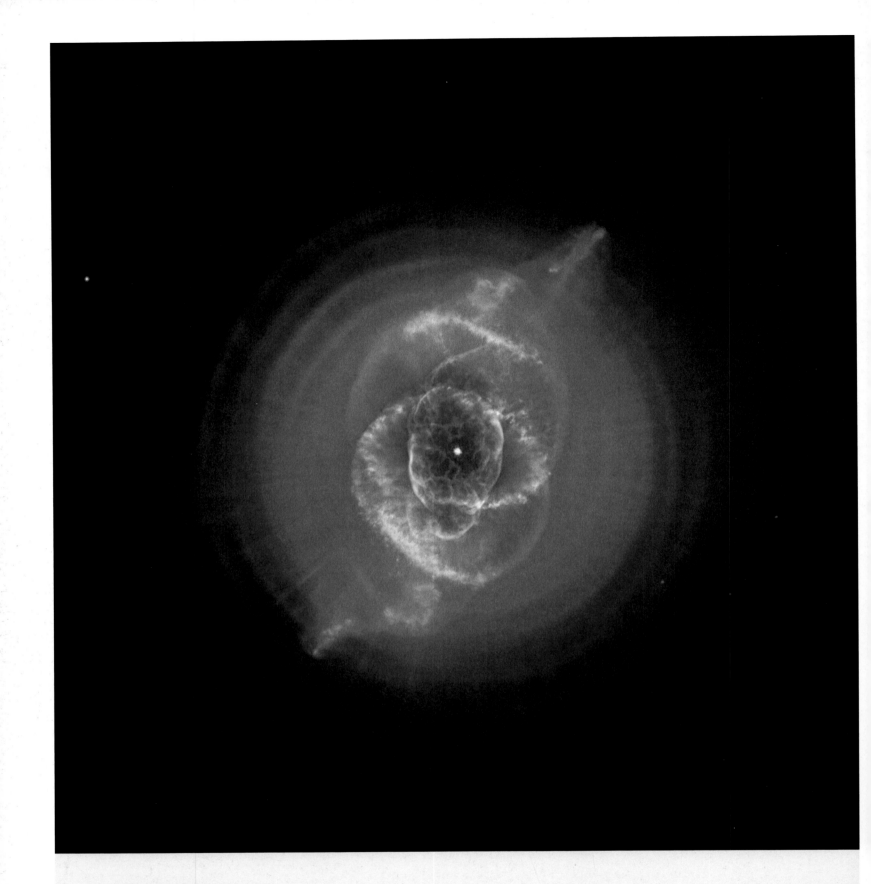

PLANETARY NEBULAE

Hubble's survey of planetary nebulae reveals surprisingly intricate, glowing patterns spun into space by aging stars: pinwheels, lawn sprinkler–style jets, elegant goblet shapes, and even some that look like a rocket engine's exhaust. These nebulae record the complex processes that happen in the final stages of a Sun-like star's evolution when it burns out and collapses to a white dwarf star. This is the Cat's Eye Nebula (NGC 6543), one of the first to be discovered. Credit: Hubble Heritage Team

The Butterfly Nebula, or Bug Nebula (NGC 6302), is roughly 3,800 light-years away, in Constellation Scorpius. What resemble dainty butterfly wings are actually roiling cauldrons of gas heated to more than 36,000 degrees Fahrenheit. The gas is tearing across space at more than 600,000 miles an hour—fast enough to travel from Earth to the Moon in 24 minutes! A dying star that was once about five times the mass of the Sun is at the center of this fury. It has ejected its envelope of gases and is now unleashing a stream of ultraviolet radiation that is making the cast-off material glow.

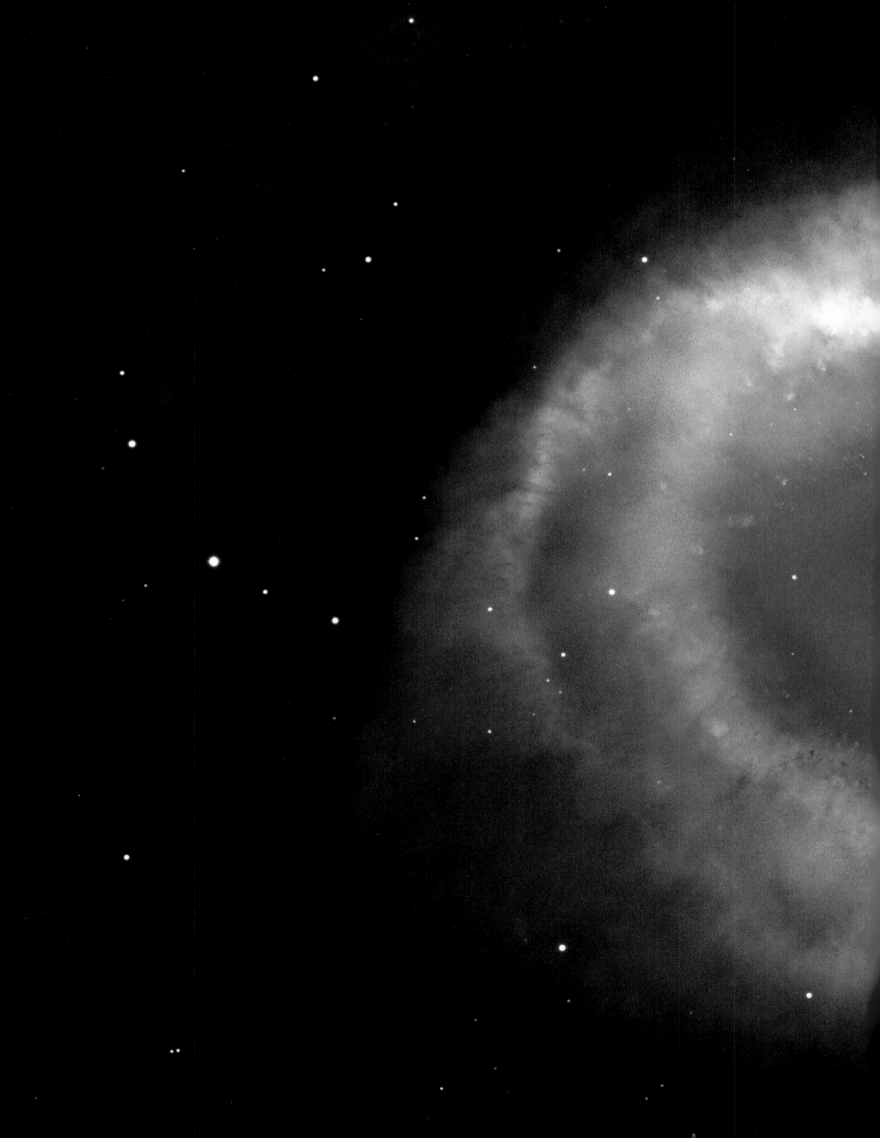

A fine web of filamentary "bicycle spoke" features is embedded in the colorful red-and-blue gas ring of the Helix Nebula, one of the nearest planetary nebulae to Earth. The Helix Nebula is a popular target of amateur astronomers and can be seen with binoculars as a ghostly, greenish cloud in the constellation Aquarius. Larger amateur telescopes can resolve the ring-shaped nebula, but only the largest ground-based telescopes can resolve the radial streaks. After careful analysis, astronomers concluded the nebula really isn't a bubble, but is a cylinder that happens to be pointed toward Earth.

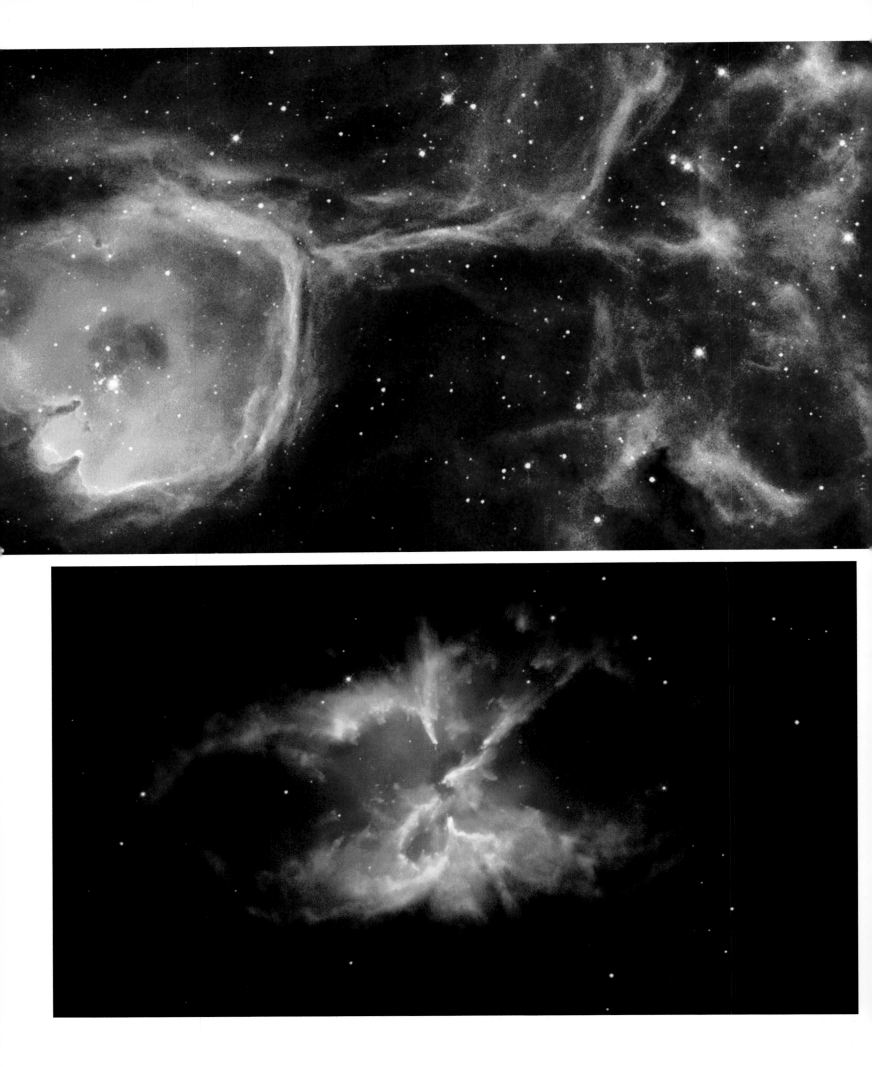

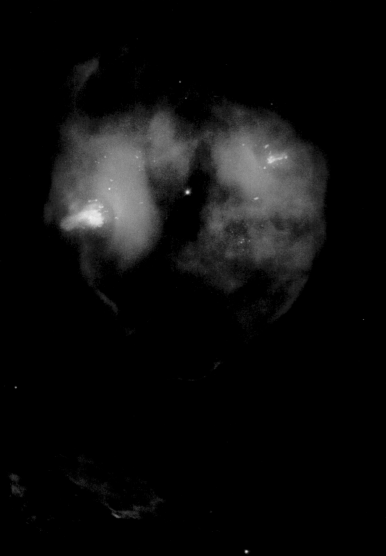

OPPOSITE, TOP: In this unusual image, Hubble captured a rare view of the celestial equivalent of a geode—a gas cavity carved by the stellar wind and intense ultraviolet radiation from a hot young star. Real geodes are baseball-size, hollow rocks that start out as bubbles in volcanic or sedimentary rock. Only when these inconspicuous round rocks are split in half by a geologist do we get a chance to appreciate the inside of the rock cavity, which is lined with crystals. Here, the transparency of the bubblelike cavity of interstellar gas and dust reveals the treasures of its interior. The object, called N44F, is being inflated by a torrent of fast-moving particles (called a "stellar wind") from an exceptionally hot star once buried inside a cold, dense cloud.

OPPOSITE, BOTTOM: Hubble has imaged striking details of the famed planetary nebula designated NGC 2818, which lies in the southern constellation of Pyxis (the Compass). Planetary nebulae fade gradually over tens of thousands of years. The hot, remnant stellar core of NGC 2818 will eventually cool off for billions of years as a white dwarf.

ABOVE: Resembling a rippling pool illuminated by underwater lights, the Egg Nebula offers astronomers a special look at the normally invisible dust shells swaddling an aging star. These dust layers, extending over one-tenth of a light-year from the star, have an onionskin structure that forms concentric rings around the star. A thicker dust belt, running almost vertically through the image, blocks off light from the central star. Twin beams of light radiate from the hidden star and illuminate the pitch-black dust, like a shining flashlight in a smoky room. The artificial "Easter-egg" colors in this image explain how the light reflected off the smoke-size dust particles and then headed toward Earth.

LEFT: Within planetary nebula NGC 2371 are the glowing remains of a Sun-like star, visible at the center. The superhot core of a former red giant, now stripped of its outer layers, it has a surface temperature of 240,000 degrees Fahrenheit. NGC 2371 lies about 4,300 light-years away, in the constellation Gemini.

This small area of the nebula near star cluster NGC 2074 (upper left) is a firestorm of raw stellar creation, perhaps triggered by a nearby supernova explosion. It lies about 170,000 light-years away near the Tarantula Nebula, in one of the most active star-forming regions in our local group of galaxies. The three-dimensional-looking image reveals dramatic ridges and valleys of dust, serpent-head "pillars of creation," and gaseous filaments glowing fiercely under torrential ultraviolet radiation. The region is on the edge of a dark molecular cloud that is an incubator for the birth of new stars.

This is a view of a turbulent cauldron of star birth called N159, 170,000 light-years away in our satellite galaxy the Large Magellanic Cloud. Torrential stellar winds from hot, newborn, massive stars within the nebula sculpt ridges, arcs, and filaments in the vast cloud, which is more than 150 light-years across. A rare type of compact, illuminated "blob" is resolved for the first time to be a butterfly-shaped or "Papillon" (French for "butterfly") Nebula, buried in the center of the maelstrom of glowing gases and dark dust. The unprecedented details of the structure of the Papillon, itself less than 2 light-years in size, are seen at right.

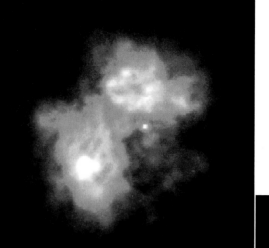

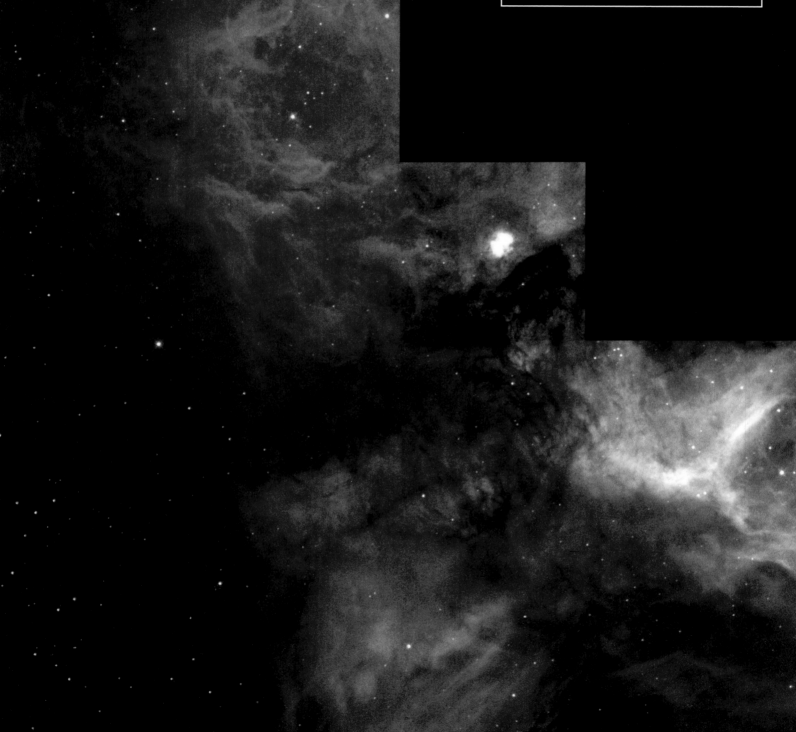

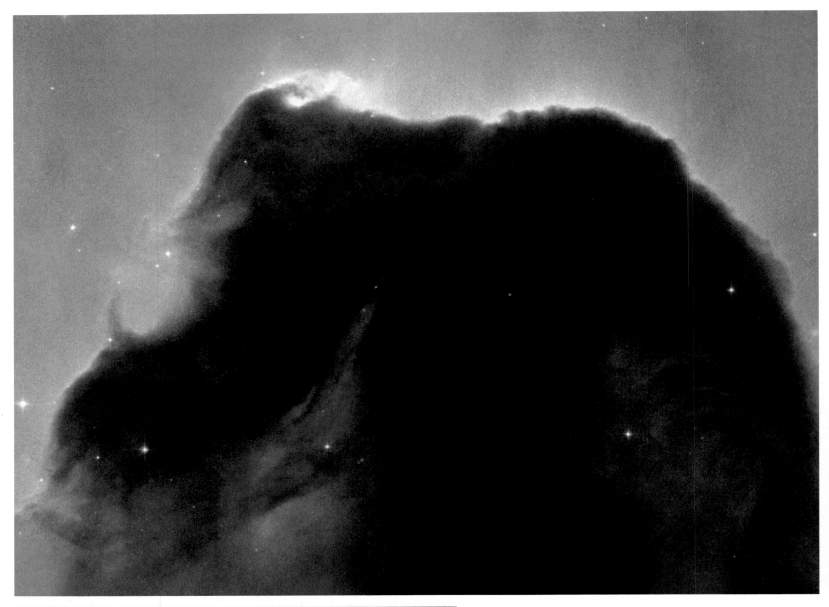

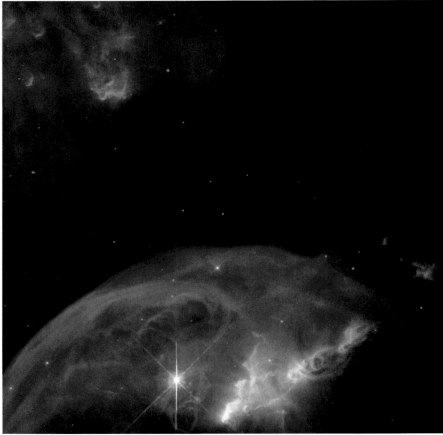

ABOVE: Rising from a sea of dust and gas like a giant sea horse, the Horsehead Nebula is one of the most photographed objects in the sky. Hubble took a close-up look at this heavenly icon, revealing the cloud's intricate structure. The Horsehead Nebula lies just south of the bright star Zeta Orionis, which is easily visible to the unaided eye as the left-hand star in the line of three that form Orion's Belt. Amateur astronomers often use the Horsehead as a test of their observation skills; it is known as one of the more difficult objects to see in an amateur-size telescope.

LEFT: The central star of the Bubble Nebula (NGC 7635) is 40 times more massive than the Sun and generates a stellar wind moving at 4 million miles (7 million kilometers) per hour. The bubble surface actually marks the leading edge of this wind's gust front, which is slowing as it plows into the denser surrounding material. At a distance of 7,100 light-years from Earth, the Bubble Nebula is located in the constellation Cassiopeia and has a diameter of 6 light-years.

Remnants from a star that exploded thousands of years ago created a celestial abstract portrait called the Pencil Nebula. Officially known as NGC 2736, the Pencil Nebula is part of the huge Vela supernova remnant, located in the southern constellation Vela. Discovered by Sir John Herschel in the 1840s, the nebula's linear appearance triggered its popular name. The nebula's shape suggests that it is part of the supernova shock wave that recently encountered a region of dense gas. It is this interaction that causes the nebula to glow, appearing like a rippled sheet.

Among the Interstellar Clouds   **59**

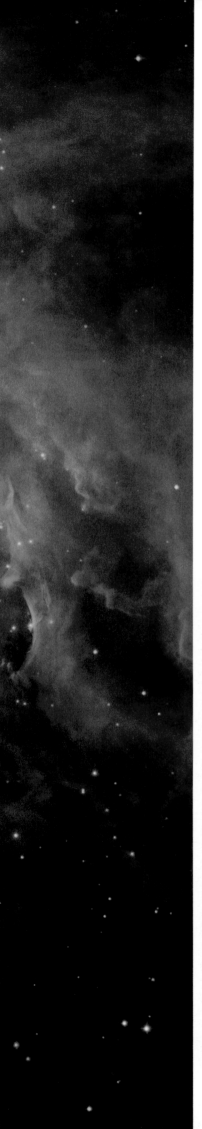

LEFT: This magnificent image from NASA's Spitzer and Hubble space telescopes shows the Orion Nebula in an explosion of infrared, ultraviolet, and visible-light colors. It was "painted" by hundreds of baby stars on a canvas of gas and dust, with intense ultraviolet light and strong stellar winds as brushes. At the heart of the nebula, in the brightest part of the image, is a group of four monstrously massive stars, collectively called the Trapezium. Located 1,500 light-years from Earth, the Orion Nebula is the brightest "star" in the sword of the Hunter constellation.

BELOW: M43, a massive star, is illuminating and sculpting the landscape of a small region with dust and gas. Astronomers call the area, which is found just above the Orion Nebula on the facing page, a miniature Orion Nebula because of its small size and the single star that shapes it. The Orion Nebula itself is much larger and has the Trapezium to carve the dust-and-gas terrain.

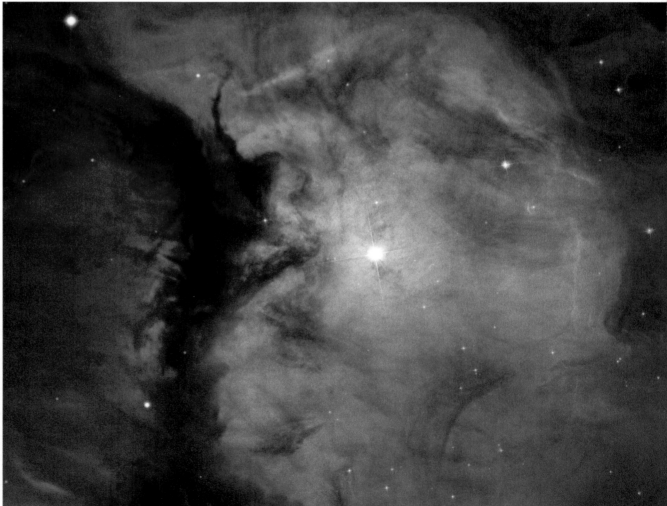

In 2007, a team of astronomers released one of the largest panoramic images ever taken with Hubble's cameras: a 50-light-year-wide view of the central region of the Carina Nebula, revealing a maelstrom of star birth and death. The Carina Nebula complex is located at a distance of roughly 7,200 light-years and lies in the constellation Carina. The immense nebula contains at least a dozen brilliant stars that are roughly estimated to be at least 50 to 100 times the mass of our Sun. Its fantasy-like landscape is sculpted by the action of outflowing winds and scorching ultraviolet radiation from these monster stars. In the process, the stars are shredding the surrounding material that is the last vestige of the giant cloud from which they were born. The most unusual and opulent inhabitant is the star Eta Carinae (see page 43), at far left, with its two billowing lobes of gas and dust.

LEFT: So-called elephant trunk pillars in the Carina Nebula (visible at the far right in the image on the previous spread, where the area seen in this detail is oriented approximately 75 degrees counterclockwise) resist being heated and eaten away by blistering ultraviolet radiation from the nebula's brightest stars. These great clouds of cold hydrogen resemble summer afternoon thunderheads. They tower above the surface of a molecular cloud on the edge of the nebula.

BELOW: This Hubble image shows the edge of a gaseous cavity within the star-forming region called NGC 3324, at the northwest corner of the Carina Nebula. The glowing nebula has been carved out by intense ultraviolet radiation and stellar winds from several hot, young stars. A cluster of extremely massive stars, located well outside this image in the center of the nebula, is responsible for the ionization of the nebula and excavation of the cavity.

LEFT: Appearing like a winged fairy-tale creature poised on a pedestal, this object is actually a billowing tower of cold gas and dust rising in the Eagle Nebula. The soaring pillar is 9.5 light-years, or about 57 trillion miles, high, about twice the distance from our Sun to the nearest star. Stars in the Eagle Nebula are born in clouds of cold hydrogen that reside in chaotic neighborhoods, where energy from young stars sculpts fantasy-like landscapes in the gas. The tower may be a giant incubator for those newborn stars. A torrent of ultraviolet light from a band of massive, hot, young stars (off the top of the image) is eroding the pillar.

OPPOSITE: A 3-light-year-long pillar in the Carina Nebula photographed in visible light, above, is bathed in the glow of light from hot, massive stars. Scorching radiation and fast winds (streams of charged particles) from these stars are sculpting the pillar and causing new stars to form within it. Streamers of gas and dust can be seen flowing off the top of the structure. The fledgling stars inside the pillar cannot be seen in the top image because they are hidden by gas and dust. Although the stars themselves are invisible, one of them is providing evidence of its existence: Thin puffs of material can be seen traveling to the left and to the right of a dark notch in the center of the pillar. The matter is part of a jet produced by a young star. Farther away, on the left, the jet is visible as a grouping of small, wispy clouds. The jet's total length is about 10 light-years.

In the image at bottom, taken in near-infrared light, the dense column and the surrounding greenish-colored gas all but disappear. Only a faint outline of the pillar remains. By penetrating the wall of gas and dust, the infrared image reveals the infant star that is probably blasting the jet.

Our own solar system resides in the Milky Way galaxy, just one of hundreds of billions of galaxies traveling through space. These galaxies are giant systems composed of tens to hundreds of billions of stars, often with vast assemblies of dust and gas. It is a testimony to their great distance from us that such giant systems appear very small to us in the

night sky. As an example, the nearest large galaxy to the Milky Way, the Andromeda galaxy, is 15 billion miles (2.5 million light-years) distant. Because light travels at a constant speed and moves across such great distances, by the time it is captured by the Hubble Space Telescope and other observatories we see the galaxies as they were many millions or billions of years ago.

In essence, Hubble is a time machine, giving us a glimpse into the distant past of our universe.

In January 1996, scientists revealed what at the time was humanity's most distant view of the universe. Known as the Hubble Deep Field image, it represented a tremendous look back in time but encompassed only a pinhole view of the visible night sky, equivalent to the amount of sky visible through a very narrow drinking straw. In this tiny field of view, about three thousand galaxies were seen.

In 2004, the Hubble Ultra Deep Field image topped the observatory's previous work and showed the oldest galaxies ever seen—an estimated ten thousand galaxies total in the field of view, with some having emitted their light about 13 billion years ago. The light Hubble is receiving today left such systems some 8.5 billion years before the Earth was even formed.

Myriad stars embedded in the heart of the nearby galaxy NGC 300 can be singled out as individual points of light in this remarkable image, despite the fact that the galaxy is 6.5 million light-years away. NGC 300 is a spiral galaxy similar to our own Milky Way galaxy. It is a member of a nearby group of galaxies known as the Sculptor Group, named for the southern constellation where the group is found. At this distance, only the brightest stars can be picked out from ground-based images. With a resolution some ten times better than that of ground-based telescopes, Hubble's Advanced Camera for Surveys (ACS) resolves many more stars in this galaxy than can be detected from the ground.

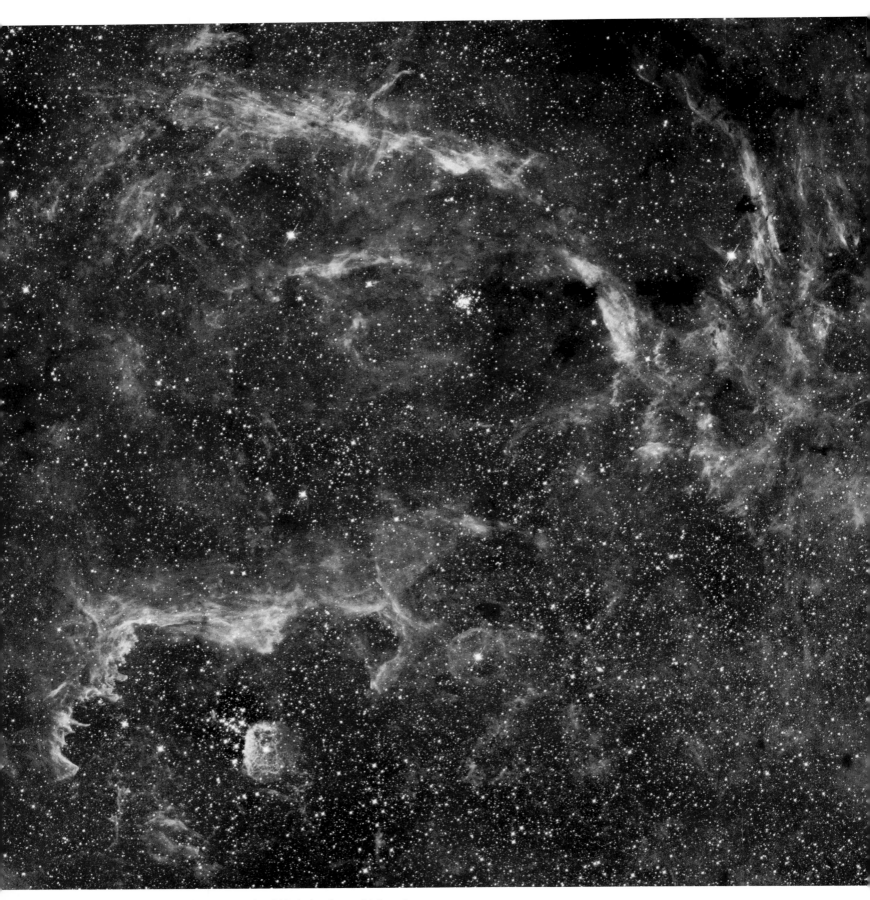

This sweeping panorama, 300 light-years in width, is the sharpest infrared picture ever made of the core of our Milky Way galaxy. It offers a nearby laboratory for how massive stars form and influence their environment in the often-violent nuclear regions of other galaxies. The galactic core is obscured in visible light by intervening dust clouds, but infrared light penetrates the dust. This view combines the sharp imaging of Hubble's Near Infrared Camera and Multi-Object Spectrometer (NICMOS) with color imagery from a previous Spitzer Space Telescope survey done with its Infrared Astronomy Camera (IRAC). The NICMOS mosaic required 144 Hubble orbits to make 2,304 exposures.

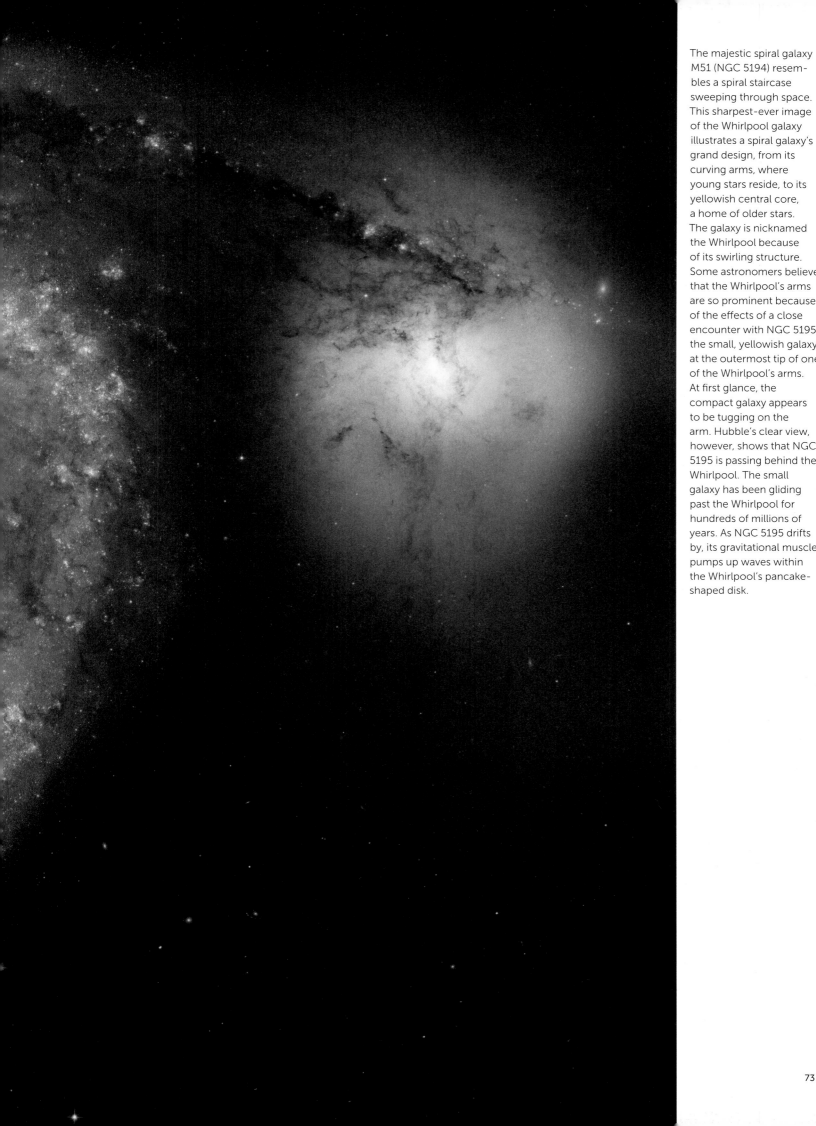

The majestic spiral galaxy
M51 (NGC 5194) resem-
bles a spiral staircase
sweeping through space.
This sharpest-ever image
of the Whirlpool galaxy
illustrates a spiral galaxy's
grand design, from its
curving arms, where
young stars reside, to its
yellowish central core,
a home of older stars.
The galaxy is nicknamed
the Whirlpool because
of its swirling structure.
Some astronomers believe
that the Whirlpool's arms
are so prominent because
of the effects of a close
encounter with NGC 5195,
the small, yellowish galaxy
at the outermost tip of one
of the Whirlpool's arms.
At first glance, the
compact galaxy appears
to be tugging on the
arm. Hubble's clear view,
however, shows that NGC
5195 is passing behind the
Whirlpool. The small
galaxy has been gliding
past the Whirlpool for
hundreds of millions of
years. As NGC 5195 drifts
by, its gravitational muscle
pumps up waves within
the Whirlpool's pancake-
shaped disk.

Hubble captures a display of starlight, glowing gas, and silhouetted dark clouds of interstellar dust in this image of the barred spiral galaxy NGC 1300. NGC 1300 is considered to be prototypical of barred spiral galaxies. Barred spirals differ from normal spiral galaxies in that the arms of the galaxy do not spiral all the way into the center, but are connected to the two ends of a straight bar of stars containing the nucleus at its center. At Hubble's resolution, a myriad of fine details, some of which have never before been seen, are visible throughout the galaxy's arms, disk, bulge, and nucleus. Blue and red supergiant stars, star clusters, and star-forming regions are well resolved across the spiral arms, and dust lanes trace out fine structures in the disk and bar. Numerous more distant galaxies are visible in the background, and are seen even through the densest regions of NGC 1300.

The sharpest image ever taken of the large "grand design" spiral galaxy M81 shows it tilted at an oblique angle to our line of sight, giving a bird's-eye view of the spiral structure. The galaxy is similar to our Milky Way, but our viewing angle provides a better picture of the typical architecture of spiral galaxies. Though the galaxy is 11.6 million light-years away, Hubble's view is so sharp, it can resolve individual stars, along with open star clusters, globular star clusters, and even glowing regions of fluorescent gas.

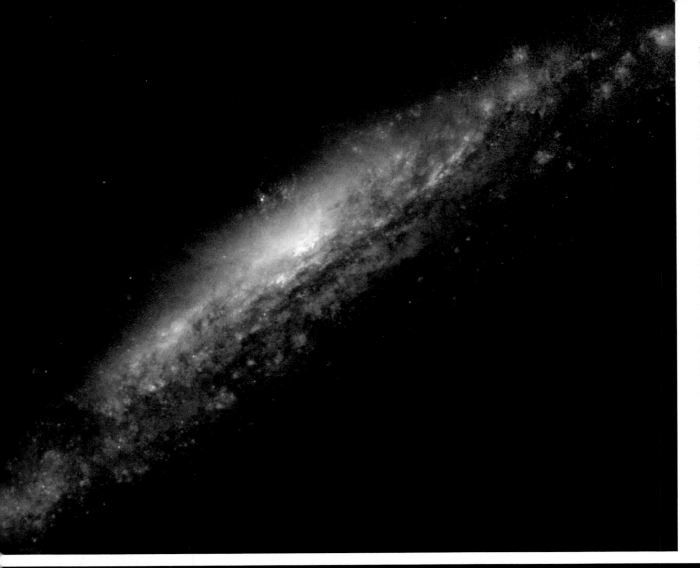

LEFT: Dramatic events are occurring within the core of the galaxy NGC 3079, where a lumpy bubble of hot gas is rising from a cauldron of glowing matter. The structure is more than 3,000 light-years wide and rises 3,500 light-years above the galaxy's disk. Astronomers suspect that the bubble is being blown by "winds" (high-speed streams of particles) released during a burst of star formation.

BELOW: The Sombrero Galaxy (M104) is a brilliant white, bulbous core encircled by the thick dust lanes that make up its spiral structure. As seen from Earth, the galaxy is tilted nearly edge on: We view it from just 6 degrees north of its equatorial plane. At a relatively bright magnitude of +8, M104 is just beyond the limit of naked-eye visibility and is easily seen through small telescopes. The Sombrero lies at the southern edge of the rich Virgo cluster of galaxies and is one of the most massive objects in that group, equivalent to 800 billion suns. The galaxy is 50,000 light-years across and is 28 million light-years from Earth.

TOP: ESO 510-G13 is an unusual galaxy, as this edge-on view of its warped dusty disk reveals. The dust and spiral arms of normal spiral galaxies, like our own Milky Way, appear flat when viewed edge on. Details of the structure of ESO 510-G13 are visible because the interstellar dust clouds that trace its disk are silhouetted from behind by light from the galaxy's bright, smooth central bulge.

BOTTOM, LEFT: A nearly perfect ring of hot, blue stars pinwheels about the yellow nucleus of an unusual galaxy known as Hoag's Object. This face-on view of the galaxy's ring of stars reveals more detail than any previous photo of this strange object. The entire galaxy is about 120,000 light-years wide, which is slightly larger than our Milky Way galaxy. The blue ring, which is dominated by clusters of young, massive stars, contrasts sharply with the yellow nucleus of mostly older stars. What appears to be a "gap" separating the two stellar populations may

actually contain some star clusters that are almost too faint to see. Curiously, an object that bears an uncanny resemblance to Hoag's Object can be seen in the gap at the one o'clock position.

BOTTOM, RIGHT: To the surprise of astronomers, NGC 4622 appears to be rotating in the opposite direction to what they expected. Hubble pictures helped astronomers determine that the galaxy may be spinning clockwise. Astronomers are puzzled by the clockwise rotation because of the direction the outer spiral arms are pointing. Most spiral galaxies have arms of gas and stars that trail behind as they turn. But this galaxy has two "leading" outer arms that point toward the direction of the galaxy's clockwise rotation. What caused this galaxy to behave differently from most galaxies? Astronomers suspect that NGC 4622 interacted with another galaxy.

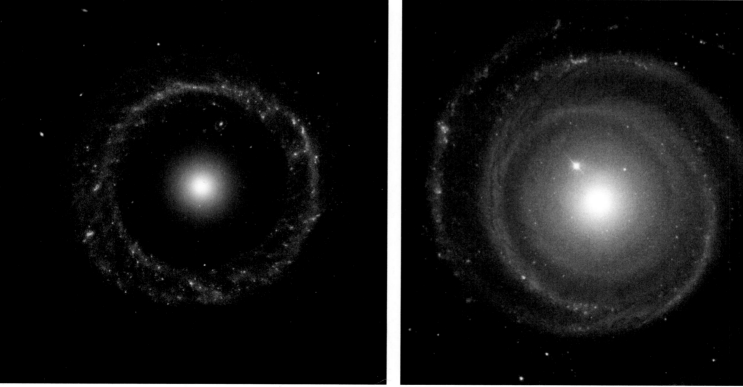

This dramatic image shows stunning details of the face-on spiral galaxy NGC 1309. NGC 1309 was home to supernova SN 2002fk, whose light reached Earth in September 2002. Observations of the galaxy taken in visible and infrared light come together in a colorful depiction of many of the galaxy's features. Bright blue areas of star formation pepper the spiral arms, while ruddy dust lanes follow the spiral structure into a yellowish central nucleus of older-population stars. The image is complemented by myriad far-off background galaxies.

Sparkling with the light from millions of newly formed young stars, NGC 1569 is one of the most active galaxies in our local neighborhood. At the nucleus of the starburst galaxy is a grouping of three giant star clusters, each containing more than a million stars. The clusters reside in a large, central cavity. The gas in the cavity has been blown out by the multitude of massive young stars that already exploded as supernovae. NGC 1569 is located 11 million light-years from Earth, in the constellation Camelopardalis.

STELLAR POPULATIONS IN M31

This image contains thousands of stars, all but a handful of which reside not in our galaxy, but in the halo of our nearest and largest galactic neighbor, the Andromeda galaxy (M31). Hubble observations allowed astronomers to reliably measure the age of the spherical halo of stars on the outskirts of the neighboring Andromeda galaxy 2.5 million light-years away. To their surprise, they have discovered that approximately one-third of the stars in Andromeda's halo formed only six to eight billion years ago. Astronomers cannot yet tell whether this was one tumultuous event or a more continual acquisition of smaller galaxies that populated the halo with stars. One possibility is that collisions destroyed the young disk of M31 and dispersed many of its stars into the halo. Alternatively, a single collision destroyed a relatively massive invading galaxy and dispersed its stars and some of Andromeda's disk stars into the halo. It is also possible that many stars formed during the collision itself. Credit: T. M. Brown (STScI)

This small galaxy, called the Sagittarius Dwarf Irregular Galaxy or "SagDIG" for short, is relatively nearby, and Hubble's sharp vision is able to reveal many thousands of individual stars within it. The brightest stars in the picture (easily distinguished by the spikes radiating from their images, produced by optical effects within the telescope) are foreground stars lying within our own Milky Way galaxy. Their distances from Earth are typically a few thousand light-years. By contrast, the numerous faint, bluish stars belong to SagDIG, which lies some 3.5 million light-years from Earth.

ABOVE: What astronomers once thought was a toddler galaxy by galactic standards may now be considered an adult. I Zwicky 18 has a youthful appearance that resembles galaxies typically found only in the early universe. Hubble has now found faint, older stars within this galaxy, suggesting that the galaxy may have formed at the same time as most other galaxies. I Zwicky 18 is classified as a dwarf irregular galaxy and is much smaller than our Milky Way galaxy. The concentrated bluish white knots embedded in the heart of the galaxy are two major starburst regions where stars are forming at a furious rate.

OPPOSITE: The disk galaxy NGC 5866 is tilted nearly edge on to our line of sight. Hubble's sharp vision reveals a crisp dust lane dividing the galaxy into two halves. The image highlights the galaxy's structure: a subtle, reddish bulge surrounding a bright nucleus, a blue disk of stars running parallel to the dust lane, and a transparent outer halo. The outer halo is dotted with numerous gravitationally bound globular clusters of nearly a million stars each. Background galaxies that are millions to billions of light-years farther away than NGC 5866 are also seen through the halo.

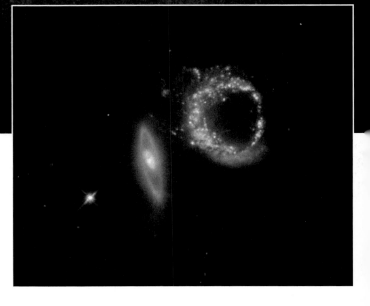

RIGHT: A pair of gravitationally interacting galaxies called Arp 147 happens to be oriented so that it appears to mark the number ten. The galaxy on the left is relatively undisturbed apart from a smooth ring of starlight. It appears nearly edge on to our line of sight. The galaxy on the right exhibits a clumpy, blue ring of intense star formation. The blue ring was most probably formed after the galaxy on the left passed through the galaxy on the right. A propagating density wave was generated at the point of impact and spread outward, and as this wave collided with material in the target galaxy that was moving inward due to the gravitational pull of the two galaxies, shocks and dense gas were produced, stimulating star formation.

In the direction of the constellation Canis Major, two spiral galaxies pass by each other like majestic ships in the night. The larger and more massive galaxy (left) is cataloged as NGC 2207 and the smaller one on the right is IC 2163. Strong tidal forces from NGC 2207 have distorted the shape of IC 2163, flinging out stars and gas into long streamers stretching out 100,000 light-years toward the right-hand edge of the image.

LEFT: Two galaxies containing a vast number of stars swing past each other in a graceful performance choreographed by gravity. The pair, known collectively as Arp 87, is one of hundreds of interacting and merging galaxies known in our nearby universe. Arp 87 is in the constellation Leo, approximately 300 million light-years away from Earth.

The smaller of these two spiral galaxies, dubbed LEDA 62867 and positioned to the left of the frame, seems to be safe for now, but will probably be swallowed by the larger spiral galaxy, NGC 6786 to the right. There is already some disturbance visible in both components. The pair is number 538 in Karachentsev's Catalog of Isolated Pairs of Galaxies. A supernova was seen to explode in the large spiral in 2004. NGC 6786 is located in the constellation Draco, about 350 million light-years away.

The galaxies of this beautiful interacting pair bear some resemblance to musical notes on a staff. Long tidal tails sweep out from the two galaxies— gas and stars were stripped out and torn away from their outer regions. The presence of these tails is the signature of an interaction. ESO 69-6 and its companion are located in the constellation Triangulum Australe, the Southern Triangle, about 650 million light-years away from Earth.

This mosaic image is the sharpest wide-angle view ever obtained of the magnificent starburst galaxy M82. The galaxy is remarkable for its bright blue disk, webs of shredded clouds, and fiery-looking plumes of glowing hydrogen blasting out of its central regions. Throughout the galaxy's center, young stars are being born ten times faster than they are inside our entire Milky Way galaxy. The young stars are crammed into tiny but massive star clusters, and these, in turn, congregate by the dozens to make the bright patches, or "starburst clumps," in the central parts of M82. The fierce galactic superwind generated by the huge concentration of young stars compresses enough gas to make millions of more stars.

This is the sharpest image taken of the merging Antennae galaxies. During the course of the collision, billions of stars are formed. The brightest and most compact of these star-birth regions are called super star clusters. The two spiral galaxies started their interaction a few hundred million years ago, making the Antennae galaxies one of the nearest and youngest examples of a pair of colliding galaxies. Nearly half of the faint objects in the Antennae image are young clusters containing tens of thousands of stars. The orange blobs to the left and right of center are the two cores of the original galaxies and consist mainly of old stars crisscrossed by filaments of dust, which appear brown in the image. The two galaxies are dotted with brilliant blue star-forming regions surrounded by glowing hydrogen gas, appearing in the image in pink.

This composite image shows the massive galaxy cluster MACS J0717.5+3745 (MACS J0717, for short), where four separate galaxy clusters have been involved in a collision—the first time such a phenomenon has been documented. Hot gas is shown in an image from the Chandra X-ray Observatory, and galaxies are shown in a Hubble optical image. The hot gas is color coded to show temperature, where the coolest gas is reddish purple, the hottest gas is blue, and the temperatures in between are purple.

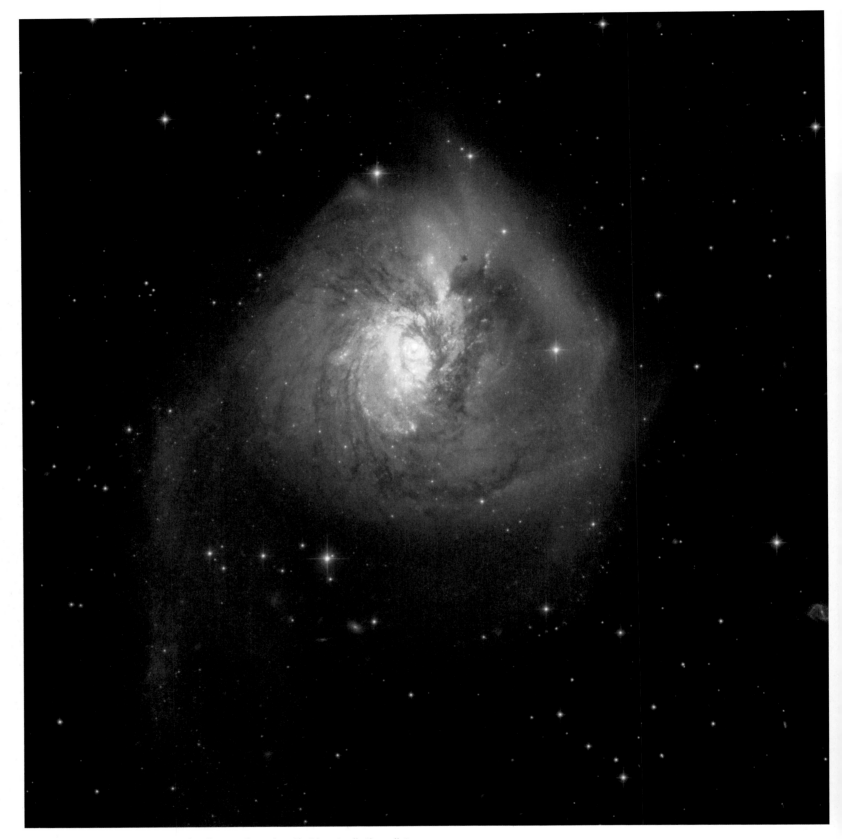

NGC 3256 is an impressive example of a peculiar galaxy that is actually the relict of a collision of two separate galaxies that took place in a distant past. The telltale signs of the collision are two extended luminous tails swirling out from the galaxy. NGC 3256 belongs to the Hydra-Centaurus supercluster complex and provides a nearby template for studying the properties of young star clusters in tidal tails. The system hides a double nucleus and a tangle of dust lanes in the central region. The tails are studded with a particularly high density of star clusters.

Like dust bunnies that lurk in corners and under beds, surprisingly complex loops and blobs of cosmic dust lie hidden in the giant elliptical galaxy NGC 1316. This image made from Hubble data reveals the dust lanes and star clusters of this giant galaxy that give evidence that it was formed from a past merger of two gas-rich galaxies. The combination of Hubble's superb spatial resolution and the sensitivity of the Advanced Camera for Surveys (ACS) instrument enabled uniquely accurate measurements of a class of red star clusters in NGC 1316. Astronomers conclude that these star clusters constitute clear evidence of the occurrence of a major collision of two spiral galaxies that merged together a few billion years ago to shape NGC 1316 as it appears today.

The monstrous elliptical galaxy M87 is the home of several trillion stars, a supermassive black hole, and 13,000 globular star clusters. M87 is the dominant galaxy at the center of the neighboring Virgo cluster of galaxies, which contains some 2,000 galaxies. Amid the smooth, yellow population of older stars, the two features that stand out most in this Hubble image of M87 are its soft blue jet and the myriad starlike globular clusters scattered throughout the image. The jet is a black hole–powered stream of subatomic particles that is accelerated to velocities near the speed of light and ejected from the core of the galaxy. As gaseous material from the center of the galaxy accretes onto the black hole, the resultant energy released produces the jet.

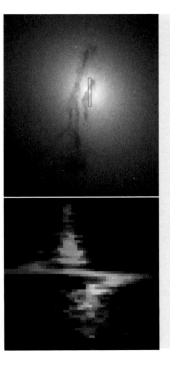

MEASURING THE MASS OF SUPERMASSIVE BLACK HOLES

Hubble probed the dense, central regions of galaxies and provided decisive evidence that supermassive black holes reside in most of them. Giant black holes are compact "monsters" weighing millions to billions of times the mass of our Sun. They cannot be observed directly, because nothing, not even light, escapes their grasp. But Hubble helped astronomers determine the masses of several black holes by measuring the velocities of material whirling around them. The speed of the orbiting material can be used to measure the mass of the black hole. The telescope's census of many galaxies revealed that a black hole's mass is dependent on the weight of its host galaxy's bulge, a spherical region consisting of stars in a galaxy's central region. The spectrographic signature, bottom left, maps the motions of gas in the region of a suspected black hole, outlined in red, top left. Credit: Holland Ford (The Johns Hopkins University)

What may first appear as a sunny-side-up egg is actually a face-on snapshot of the small spiral galaxy NGC 7742 in the constellation Pegasus. NGC 7742 is known to be a Seyfert 2 active galaxy, a type of galaxy that is probably powered by a black hole residing in its core. The core of NGC 7742 is the large yellow "yolk" in the center of the image. The lumpy, thick ring about 3,000 light-years from the core is an area of active star birth. Tightly wound spiral arms also are faintly visible. Surrounding the inner ring is a wispy band of material, which is probably the remains of a once very active stellar breeding ground.

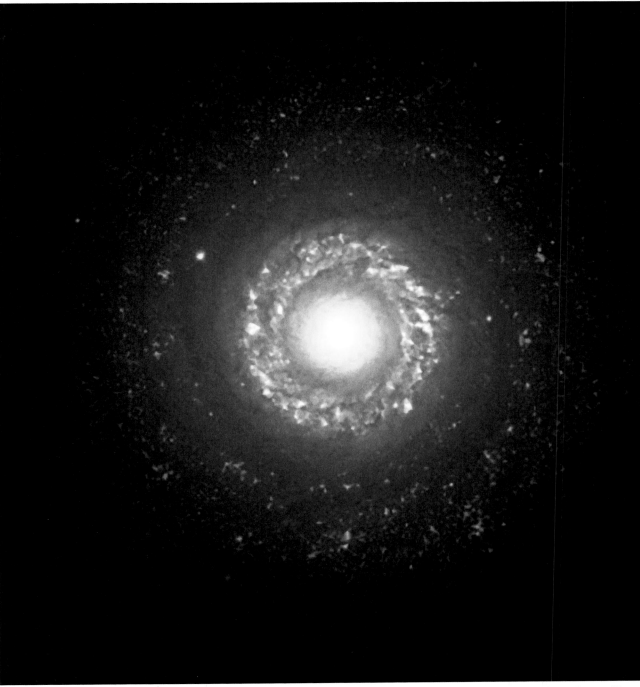

ABOVE: NGC 7174, a small spiral galaxy (middle right), is caught between two elliptical galaxies—NGC 7173 (middle left) and NGC 7176 (lower right)—in a cluster called the Hickson Compact Group 90. The elliptical galaxies are stretching, and will eventually swallow the smaller galaxy, which appears as though it is being ripped apart by its close neighbors. The galaxies are experiencing a strong gravitational interaction, and as a result, a significant number of stars have been ripped away from their home galaxies, forming a tenuous luminous component in the galaxy group.

OPPOSITE: A clash among members of a famous galaxy quintet reveals an assortment of stars across a wide color range, from young blue stars to aging red stars. This portrait of Stephan's Quintet, also known as Hickson Compact Group 92, is a group of five galaxies. The name, however, is a bit of a misnomer. Studies have shown that group member NGC 7320, at upper left, is actually a foreground galaxy about seven times closer to Earth than the rest of the group. The rest of the group comprises NGC 7319 (top right), NGC 7318A (center), and NGC 7317 (bottom left).

The Coma Cluster of Galaxies is one of the densest known galaxy collections in the universe. The Hubble's Advanced Camera for Surveys (ACS) viewed a large portion of the cluster, spanning several million light-years across. The entire cluster contains thousands of galaxies in a spherical shape more than 20 million light-years in diameter. Also known as Abell 1656, the Coma Cluster is more than 300 million light-years away. The cluster, named after its parent constellation, Coma Berenices, is near the Milky Way's north pole. This places the Coma Cluster in an area unobscured by dust and gas from the plane of the Milky Way, and easily visible by Earth viewers.

PRECISE CALIBRATION OF COSMIC DISTANCE SCALE

The Hubble image at left shows a class of pulsating star called a Cepheid variable in galaxy M100. Though rare, these pulsations can be used to make a precise measurement of the star's distance from Earth. Based on the Hubble observation, the distance to M100, above, has been measured accurately as 56 million light-years, making it one of the farthest objects where intergalactic distances have been determined precisely. Credit: Wendy L. Freedman, Observatories of the Carnegie Institution of Washington

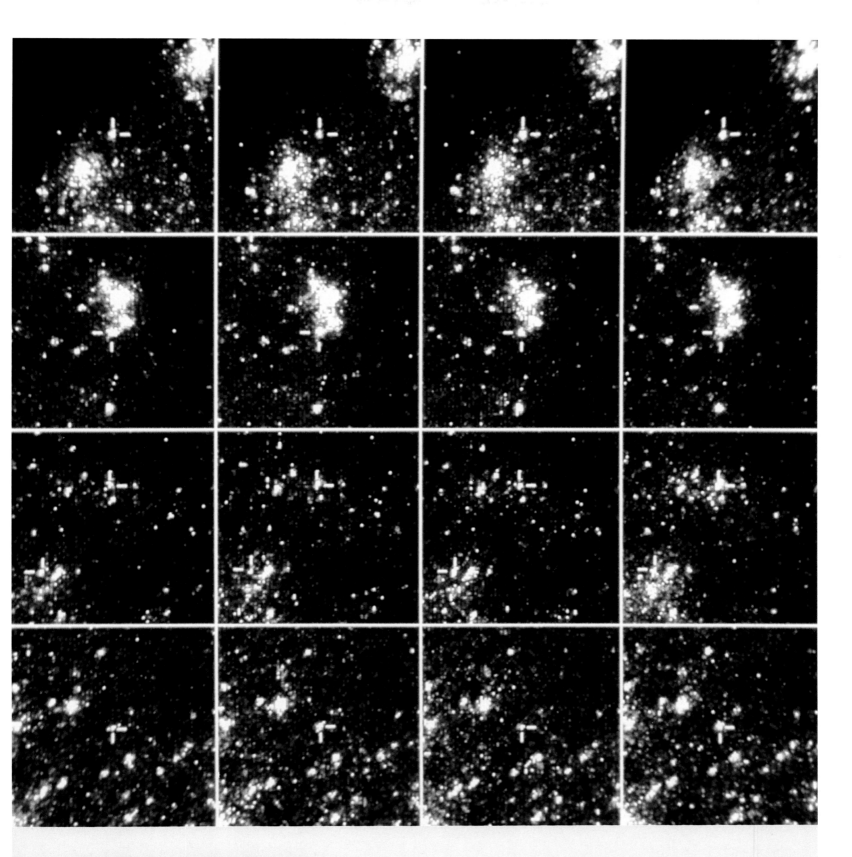

PRECISE MEASURE OF HUBBLE CONSTANT

Hubble observations allowed astronomers to calculate a precise expansion rate for the universe, a value called the Hubble constant. In May 1999 a team of astronomers obtained a value for the Hubble constant by measuring the distances to nearly two dozen galaxies, some as far as 65 million light-years from Earth. The team used the Hubble constant to determine that the universe is about 13 billion years old. In 2003, Hubble observations combined with data from NASA's Wilkinson Microwave Anisotropy Probe (WMAP) refined the value of the universe's expansion rate to a precision of 3 percent. Credit: Wendy L. Freedman, Observatories of the Carnegie Institution of Washington

This Hubble composite image shows a ghostly "ring" depicting a computer model of dark matter in the galaxy cluster Cl 0024+17 overlaid on a Hubble picture of the cluster. The ringlike structure is evident in the blue map of the cluster's dark-matter distribution. The computer model was derived from Hubble observations of how the gravity of the cluster distorts the light of more distant galaxies. The ring is one of the strongest pieces of evidence to date for the existence of dark matter, an unknown substance that pervades the universe. Astronomers suggest that the dark-matter ring was produced from a collision between two gigantic clusters. Dark matter makes up the bulk of the universe's material and is believed to make up the underlying structure of the cosmos.

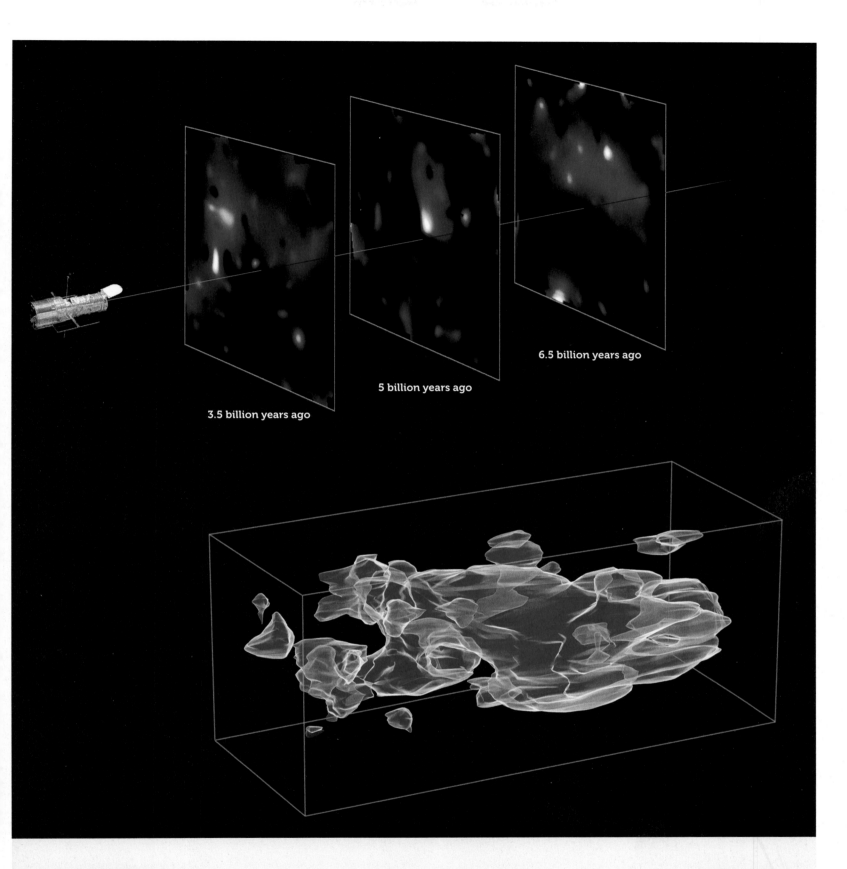

3.5 billion years ago

5 billion years ago

6.5 billion years ago

DARK-MATTER MAP

Hubble observations allowed astronomers to make a three-dimensional map that offers a look at the weblike, large-scale distribution of dark matter, an invisible form of matter that accounts for most of the universe's mass. The map reveals a loose network of dark-matter filaments, gradually collapsing under the pull of gravity and growing clumpier over time. This confirms theories of how structure formed in our evolving universe, which has transitioned from a comparatively smooth distribution of matter at the time of the Big Bang. The dark matter filaments began to form first and provided underlying scaffolding for the subsequent construction of stars and galaxies from ordinary matter. Credit: R. Massey (California Institute of Technology)

This is the first picture of a group of five starlike images of a single distant quasar. The multiple-image effect seen in the Hubble picture is produced by a process called gravitational lensing, in which the gravitational field of a massive object (in this case, a cluster of galaxies) bends and amplifies light from an object (in this case, a quasar) farther behind it.

QUASAR HOST-GALAXY MORPHOLOGY

Quasars are compact light sources resembling stars, yet they are billions of light-years away and several hundred billion times brighter than normal stars. Hubble found that quasars reside in a variety of galaxies, from normal to highly disturbed. This put the final puzzle piece in place after thirty years of observations with ground-based telescopes, following the discovery of quasars in the early 1960s. Astronomers believe that a quasar turns on when a massive black hole at the nucleus of a galaxy feeds on gas and stars. As the matter falls into the black hole, intense radiation is emitted. Quasars were more abundant in the early universe, when galaxies were more prone to interact and thereby provide the gas for "turning on" a quasar. The image at left shows one quasar host galaxy. Credit: John Bahcall (Institute for Advanced Study, Princeton), Mike Disney (University of Wales)

EVIDENCE FOR DARK ENERGY

Dark energy is a mysterious repulsive force that causes the universe to expand at an increasing rate. The Hubble Space Telescope uncovered evidence for dark energy by studying the expansion rate of the universe, as measured by the light of distant supernovae. Hubble discovered that dark energy is not a new constituent of space, but rather has been present for most of the universe's history. This picture of dark energy is consistent with Albert Einstein's prediction of nearly a century ago that a repulsive form of gravity emanates from empty space. The image above shows a supernova blast 10 billion light-years from Earth. Credit: A. Riess (The Johns Hopkins University)

BELOW: A massive cluster of yellowish galaxies, seemingly caught in a red and blue spiderweb of eerily distorted background galaxies, makes for a spellbinding picture. To create this unprecedented image of the cosmos, Hubble peered straight through the center of one of the most massive galaxy clusters known, called Abell 1689. The gravity of the cluster's trillion stars—plus dark matter—acts as a 2-million-light-year-wide "lens" in space. This "gravitational lens" bends and magnifies the light of the galaxies located far behind it. Some of the faintest objects in the picture are probably more than 13 billion light-years away.

OPPOSITE: Prior to the release of the Hubble Ultra Deep Field in 2004, the Hubble Deep Field of 1996 provided mankind's deepest, most detailed visible view of the universe. Representing a narrow "keyhole" view stretching to the visible horizon of the universe, the Hubble Deep Field image covers a speck of the sky only about the width of a dime 75 feet away. Though the field is a very small sample of the heavens, it is considered representative of the distribution of galaxies in space, because the universe, statistically, looks largely the same in all directions. Looking through this keyhole, Hubble uncovered a bewildering assortment of at least 1,500 galaxies at various stages of evolution.

HUBBLE ULTRA DEEP FIELD (OVERLEAF)

Hubble provided the deepest portrait of the visible universe ever achieved, taking astronomers to within a few hundred million years of the Big Bang itself. Called the Hubble Ultra Deep Field (HUDF), the million-second-long exposure found young galaxies that looked different from mature galaxies in the local universe. These different-appearing galaxies hint at a complex story of galaxy evolution over time. In ground-based photographs this patch of sky, no larger than the area visible through an eight-foot-long soda straw, is largely empty. Credit: S. Beckwith (STScI) and the HUDF Team

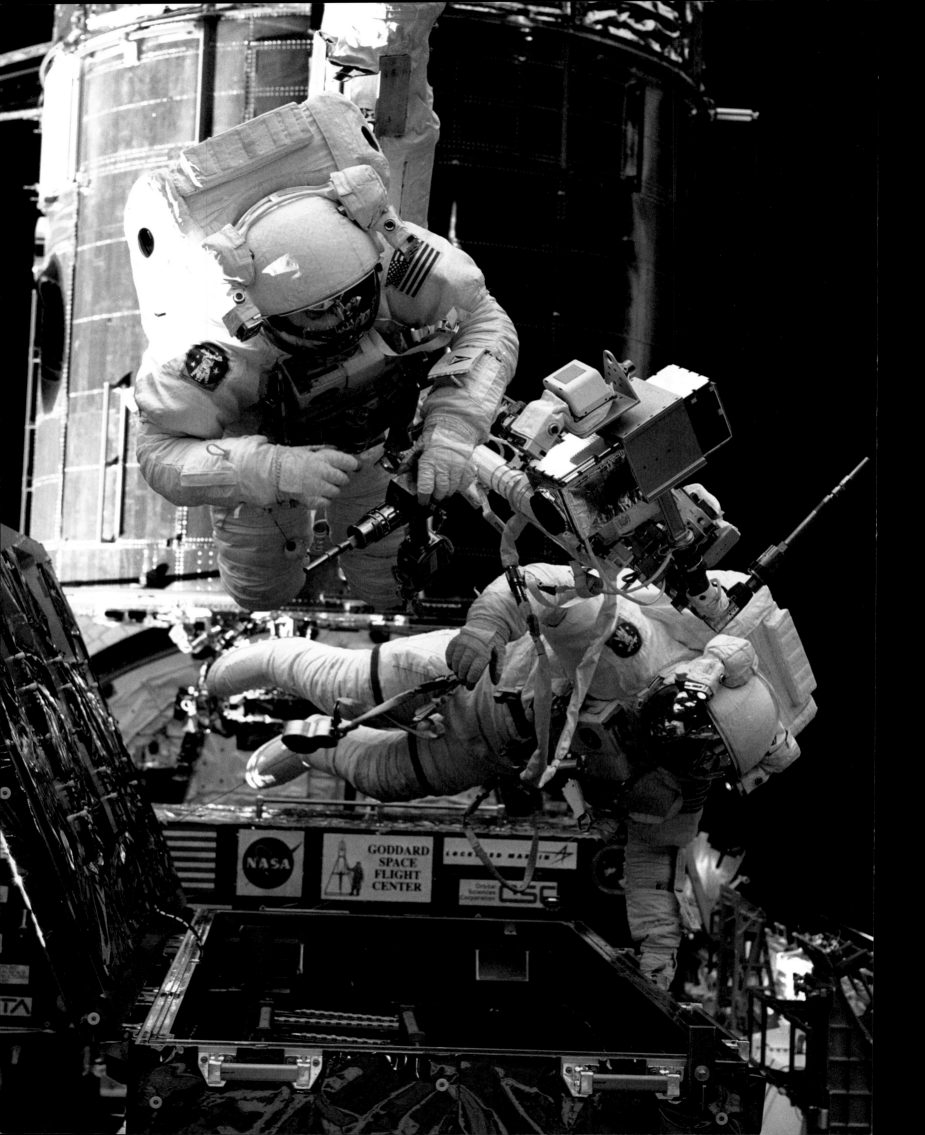

The unparalleled success and ongoing achievements of the Hubble Space Telescope are due, at least in part, to its unique design, which allows astronauts to replace and repair vital components. While the observatory is one of NASA's most stunning achievements, without human intervention and mission-saving upgrades Hubble's story would have ended much sooner and quite differently after its release in 1990.

With the benefit of two decades of hindsight, it is easy to forget the challenges faced by NASA and Hubble's many scientists and engineers from 1990 to 1993. Because of an almost unimaginably tiny error during the grinding of the observatory's primary mirror, the initial images Hubble captured were blurry. Instead of unveiling secrets of the cosmos, Hubble tested NASA's resolve. However, in a series of daring spacewalks by the STS-61 crew in December 1993, Hubble received its corrective lenses and began to see the universe as astronomers and engineers had hoped.

Over the past twenty years, astronauts have made four additional onorbit house calls, in 1997, 1999, 2002, and 2009, to keep the observatory healthy, upgrade its capabilities, and extend Hubble's life span. Collectively these servicing missions account for 166 hours and 6 minutes of extravehicular activity—spacewalks—to repair and upgrade the telescope. Acting often with the precision of surgeons, astronauts have replaced everything from gyroscopes and batteries to the telescope's main cameras and other instruments—all of which have been necessary to help Hubble maintain the power and vision that has changed the way we see our universe.

This important collaboration between humans and machines is stunningly illustrated in this chapter and captured through the observations of the astronauts who helped change the outcome of Hubble's story. These periodic checkups and hardware replacement missions have ensured that Hubble will continue to make amazing discoveries for years to come.

Gearing up with tools for the first spacewalk of the second Hubble servicing mission are astronauts Steven L. Smith (left) and Mark C. Lee. They were among four STS-82 crew members who shared five two-member spacewalking work sessions during the flight. The photograph was made from inside the space shuttle *Discovery*'s cabin.

Mission STS-31

The Hubble Space Telescope was carried into space on the five-day shuttle mission STS-31, which launched on April 24, 1990, and released the satellite into orbit a day later. The space telescope carried five scientific instruments: the Wide Field and Planetary Camera, the Goddard High Resolution Spectrograph, the High Speed Photometer, the Faint Object Camera, and the Faint Object Spectrograph. The crew of the space shuttle *Discovery* included Loren J. Shriver, commander; Charles F. Bolden, Jr., pilot; and Steven A. Hawley, Bruce McCandless II, and Kathryn D. Sullivan, mission specialists.

Dr. Kathryn D. Sullivan and I had been selected for our prior EVA experience and our willingness to undertake a "Flight Day 2" EVA if necessary. Col. Loren J. Shriver, USAF, assumed command of our crew following the loss of Challenger; Dr. Steven A. Hawley was the manipulator operator, charged with "not breaking" the telescope while extracting it from the payload bay with only inches of clearance. Col. Charles F. Bolden, Jr., USMC, now the twelfth NASA Administrator, completed the crew as pilot and "executive officer."

When finally launched, we pushed the orbiter almost to its limits, reaching the shuttle record altitude of 330 nautical miles to give Hubble the longest lifetime before needing reboost. All went flawlessly until the deployment of the second solar array. It unfurled only about 18 inches and stopped. Repeated ground commanding caused only incremental movements. Kathy and I went below to prepare for the anticipated manual EVA deployment of the balky array. With Charlie assisting, we suited up, got into the airlock, and ultimately depressurized to 5 psi. In the meantime payload systems trouble-shooters had isolated the problem to the blanket Tension Monitoring Module, and generated a software "patch" to by-pass it—which solved the solar array deployment problem.

In short order thereafter Steve released the HST, Loren backed us away from it, and Kathy and I returned to the intra-vehicular world. Our last good view of it was looking down, with the HST clearly silhouetted against a brilliant white Chilean salt pan. At the time Kathy and I were disappointed to have "come so close" to an EVA, yet be denied the actual opportunity. In retrospect, however, it was probably better that our "EVA fingerprints" were not all over the telescope when the difficulties with the primary mirror and the solar arrays were later encountered. —BRUCE McCANDLESS II

ABOVE: STS-31 crew members wearing mission T-shirts pose for their onboard crew portrait on the middeck of *Discovery*. Left to right are Charles F. Bolden, Jr. (top left), Loren J. Shriver, Kathryn D. Sullivan, Bruce McCandless II, and Steven A. Hawley.

OPPOSITE: MS Bruce McCandless II (left, in spacesuit) and MS Kathryn D. Sullivan make a practice spacewalk in Johnson Space Center's Weightless Environment Training Facility during training for the STS-31 mission. Scuba-equipped divers monitor the crew members during this simulated extravehicular activity. The duo trained for spacewalks, but none proved necessary on the mission.

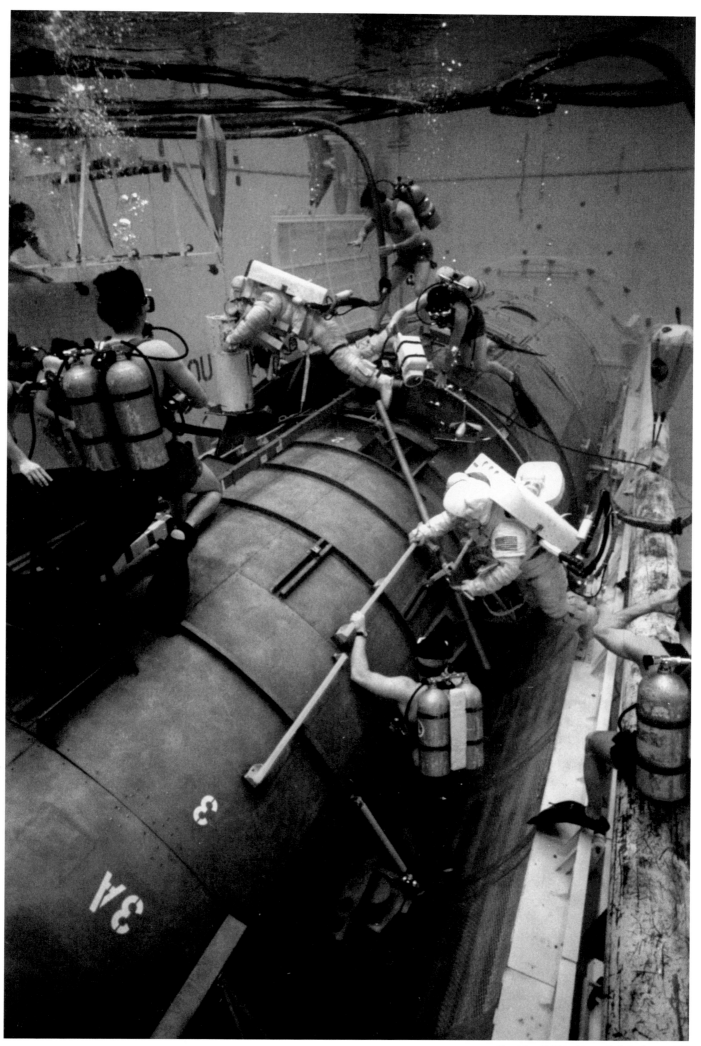

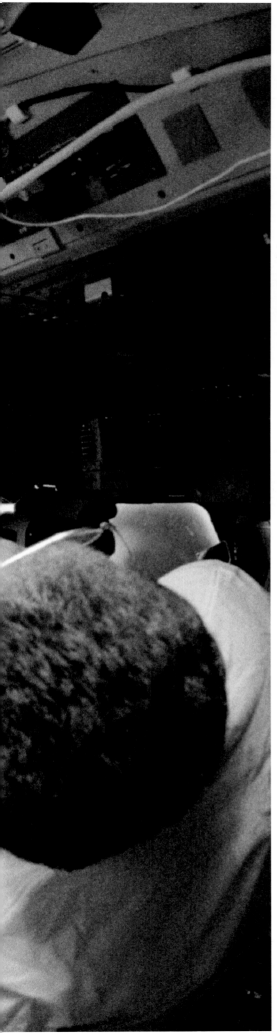

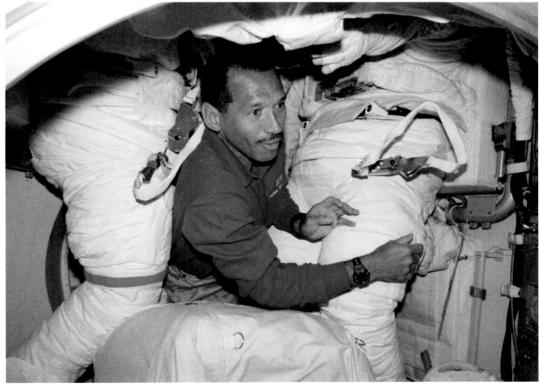

LEFT: On the aft flight deck of *Discovery*, Loren J. Shriver, Steven A. Hawley, and Bruce McCandless II (from front to back) are looking up through the overhead windows at the Hubble Space Telescope, which is still attached to the Remote Manipulator System.

TOP: In this view of space shuttle *Discovery*'s flight deck, Kathryn D. Sullivan, holding a camera, is positioned between the commander's and pilot's stations above the center console. Commander Loren J. Shriver leans against the aft flight deck onorbit station as he reviews a checklist. The photograph was taken with a fish-eye lens.

ABOVE: Pilot Charles F. Bolden, Jr., checks two spacesuits in *Discovery*'s airlock. Astronauts Bruce McCandless II and Kathryn D. Sullivan suited up for a spacewalk to free a jammed solar array, but the problem was solved remotely and they remained aboard the space shuttle.

Mission STS-61

The first Hubble servicing mission, STS-61, was launched on December 2, 1993, and lasted eleven days. In five spacewalks, the mission fulfilled all of its objectives, primarily replacing the telescope's Wide Field and Planetary Camera with the Wide Field and Planetary Camera 2 and replacing the High Speed Photometer with COSTAR, a corrective optics system. These fixes made it possible for the space telescope to serve its original purpose. STS-61 also performed upgrades and maintenance and confirmed the practicality of a satellite that could be serviced in space. The crew of the space shuttle *Endeavour* included Richard O. Covey, commander; Kenneth D. Bowersox, pilot; F. Story Musgrave, payload commander; and Thomas D. Akers, Jeffrey A. Hoffman, Claude Nicollier, and Kathryn C. Thornton, mission specialists.

The most outstanding memory I have of STS-61 is that our rescue of the initially crippled Hubble Space Telescope actually worked! Given the incredible success that Hubble has had over the years and the remarkable accomplishments of the four successive refurbishment missions, it is hard to remember the doubts that so many people had as to whether NASA had "bitten off more than it could chew" in attempting a mission as complex as repairing Hubble's optical flaw, to say nothing of the dozen other tasks we were to attempt. When Story Musgrave and I came in from the fifth and final EVA of the mission, knowing that we had accomplished all the tasks we had been given, the entire crew was both elated and relieved. The image of Hubble after we set it free back into its own orbit has stayed with me to this day. We altered our own orbit so that Hubble receded from us slowly to the west. At each sunrise, it was lit up as a glorious "Morning Star in the West." (The real Morning Star is always in the east!) Every orbit, it got smaller and smaller, but I always kept a lookout for it, right up to our deorbit burn. Of course, we had no way at the time of knowing if the corrective optics we had installed would really work. My other indelible memory of the mission came a few weeks after we were back on Earth, during the wee hours following New Year's Eve, when I received a call from an astronomer friend at the Space Telescope Science Institute asking if I had any champagne in the refrigerator. It turned out we had part of a bottle left over from our party, and my friend said, "Well pour a glass and raise a toast to Hubble, because we just got in the first image, and it looks like your mission was a success. Hubble is back in business!" —JEFFREY A. HOFFMAN

ABOVE: With the Hubble Space Telescope berthed in *Endeavour*'s cargo bay, crew members for the STS-61 mission pause for a crew portrait on the flight deck. Left to right are F. Story Musgrave, Richard O. Covey, Claude Nicollier, Jeffrey A. Hoffman, Kenneth D. Bowersox, Kathryn C. Thornton, and Thomas D. Akers.

OPPOSITE: Astronaut F. Story Musgrave, anchored on the end of the Remote Manipulator System arm, prepares to be elevated to the top of the towering Hubble Space Telescope to install protective covers on the magnetometers. Astronaut Jeffrey A. Hoffman (bottom of frame) assisted Musgrave with final servicing tasks on the telescope.

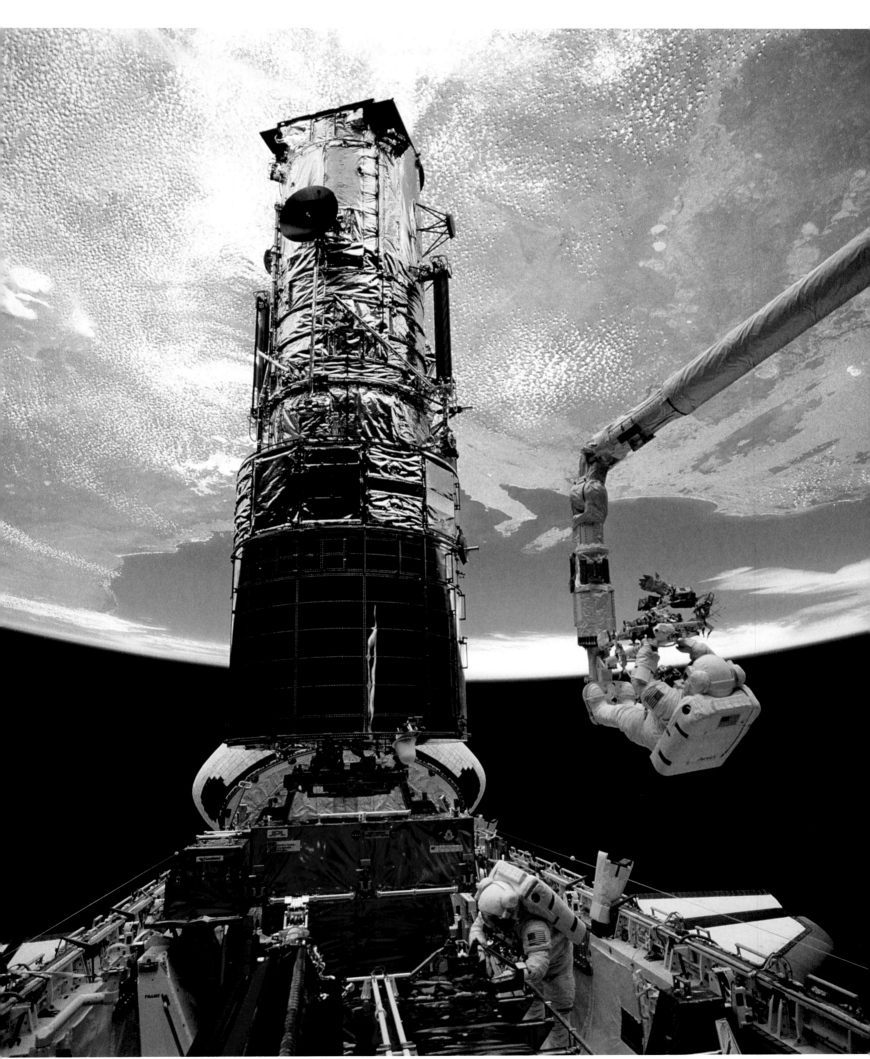

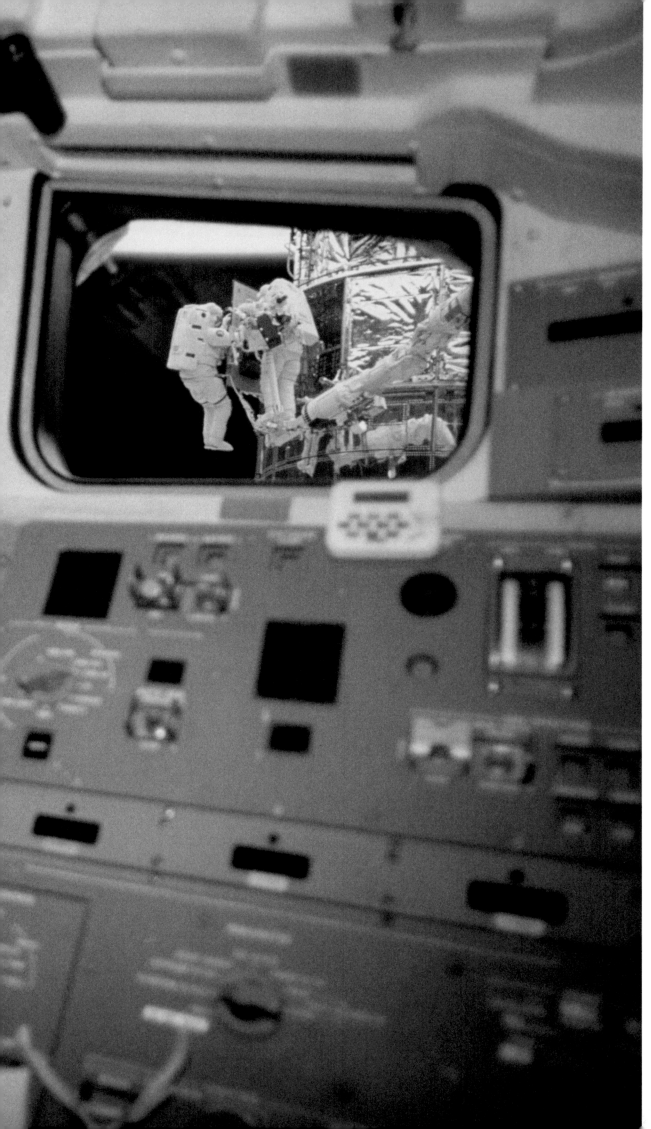

Swiss astronaut Claude Nicollier is stationed on *Endeavour*'s flight deck during one of the five spacewalks on the mission. The controls for the Remote Manipulator System are left of frame center. Two space-walkers can be seen through the aft windows.

Mission STS-82

The second Hubble servicing mission, STS-82, was launched on February 11, 1997, and lasted ten days. The crew made five spacewalks and replaced the Goddard High Resolution Spectrograph and the Faint Object Spectrograph with the Space Telescope Imaging Spectrograph and the Near Infrared Camera and Multi-Object Spectrometer (NICMOS), in addition to performing upgrades and maintenance. The crew of the space shuttle *Discovery* included Kenneth D. Bowersox, commander; Scott J. Horowitz, pilot; and Gregory J. Harbaugh, Steven A. Hawley, Mark C. Lee, Steven L. Smith, and Joseph R. Tanner, mission specialists.

STS-82 is the only mission that I flew that launched on the first attempt. The launch time resulted in our flying over the United States at night and I remember one pass that was clear from coast-to-coast. I was able to navigate my way across the country by the lights of all the cities and towns visible below. I'll never forget seeing HST as we approached and mentally comparing its appearance to my memory of how it looked when we released it seven years before. As we approached, we could see the effects of long-duration spaceflight. After we captured HST and had a chance to assess it carefully, we could see the weathered thermal protection blankets and the hole in the high gain antenna caused by a collision with a micrometeorite or orbital debris. Multiple, similar collisions had peppered the solar arrays with small holes. The handrails, which were originally a bright yellow, had a scorched, brownish tint. HST was definitely showing its age. Still, what we saw was superficial, and after our mission HST had two new instruments, a high-speed data recorder, assorted other replacement components, and a higher orbit. The real beauty of HST was the technology that enabled it to continue to be a state-of-the-art observatory. —STEVEN A. HAWLEY

ABOVE: The crew poses for a traditional in-flight portrait following completion of five spacewalks to service the Hubble Space Telescope. In front, left to right, are Joseph R. Tanner, Mark C. Lee, and Gregory J. Harbaugh. Behind them, left to right, are Steven A. Hawley, Kenneth D. Bowersox, and Scott J. Horowitz. At the very back is astronaut Steven L. Smith.

OPPOSITE: This wide shot of the Hubble Space Telescope in *Discovery*'s cargo bay, with a backdrop of Australia, was taken during the third space-walk to service the orbiting observatory. Astronaut Steven L. Smith (left of center) works near the foot restraint of the Remote Manipulator System.

BELOW: Secured to space shuttle *Discovery*'s cargo bay, the Hubble Space Telescope is seen against the backdrop of Earth's limb and atmosphere.

OPPOSITE, TOP: Astronaut Joseph R. Tanner is surrounded by a backdrop of Earth's limb and a sunburst effect in this 35mm frame exposed by astronaut Gregory J. Harbaugh, his fellow spacewalker. Harbaugh's torso is reflected in Tanner's helmet visor. A checklist of tasks is attached to Tanner's left arm for quick reference.

OPPOSITE, BOTTOM: The sun sets on the space shuttle *Discovery*'s almost empty cargo bay at the successful conclusion of the mission, as the seven astronauts inside the crew cabin approach one of the final mission chores— that of closing the cargo bay doors.

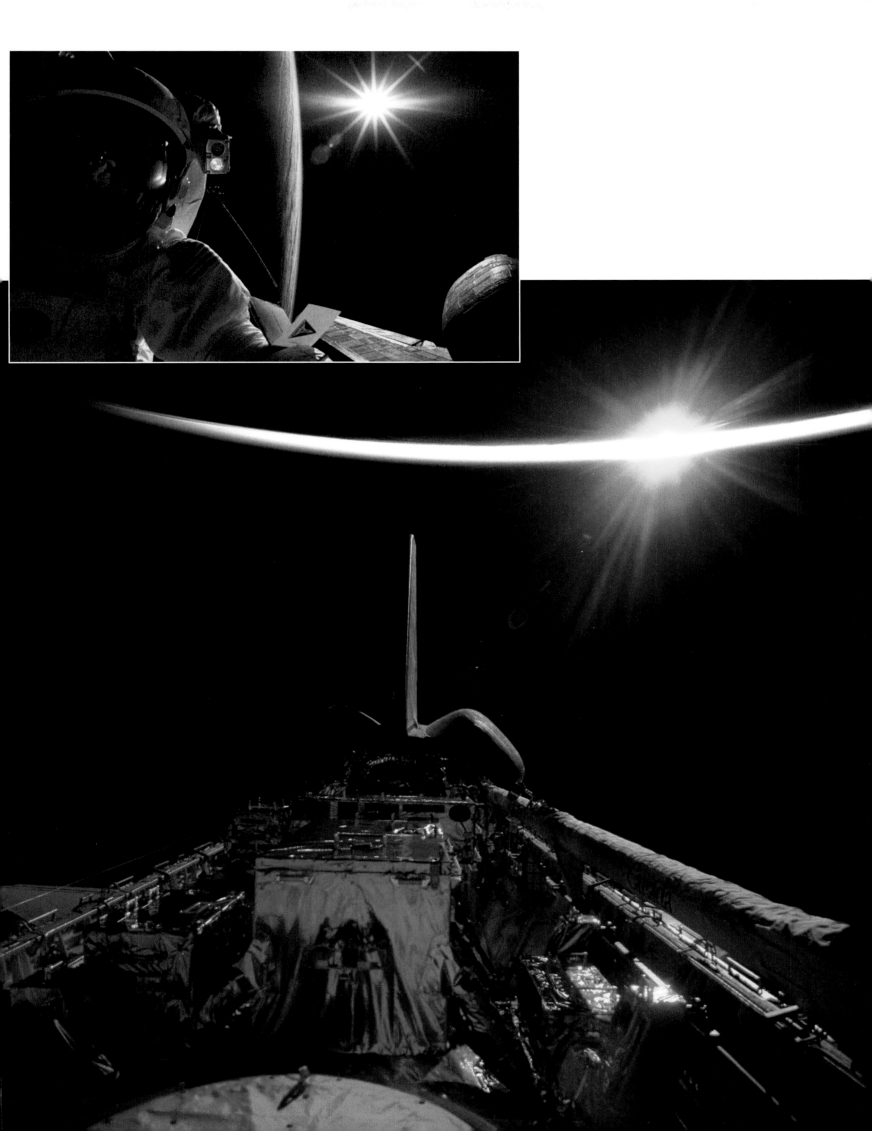

Mission STS-103

The third Hubble servicing mission, STS-103, was launched on December 19, 1999, and lasted eight days. In three spacewalks, the crew replaced the space telescope's gyroscopes and performed upgrades and maintenance. The crew of the space shuttle *Discovery* included Curtis L. Brown, commander; Scott J. Kelly, pilot; and Jean-François Clervoy, C. Michael Foale, John M. Grunsfeld, Claude Nicollier, and Steven L. Smith, mission specialists.

The four-man EVA crew of Steve Smith, Mike Foale, Claude Nicollier, and I began training for the third servicing mission to Hubble, STS-103, in 1997. We had an ambitious mission with a record six space-walks planned, including new gyroscopes, two new cooling systems, a new computer, solid-state memory, a fine-guidance sensor, a new scientific instrument, and new outer insulation for the telescope.

This grand mission was not to be. By early 1999 the Hubble was down to the minimum number of gyros to do science. NASA decided to do an emergency trip to Hubble, and our mission was cut to four spacewalks. Even then, due to a problem with the previous space-shuttle flight, our mission, now commanded by Curt Brown, with pilot Scott Kelly and flight engineer Jean-François Clervoy, was delayed. We finally made it to orbit in late December 1999, forcing us to cut our number of space-walks to three in order to beat the end of the millennium and the Y2K bugs that might result. It seemed almost like a dream when I finally reached out and touched Hubble, the Holy Grail to an astronomer/astronaut, on my first spacewalk.

Due to a failure of one of the science-data transmitters I had to perform a replacement of the failed unit. It had never been designed to be changed out onorbit. Some of the connectors are called

sub-miniature assemblies due to their tiny size. The repair of this radio set the stage for doing ever more challenging repairs on each of the following missions.

In the end we performed three space-walks of over eight hours duration each and completed the installation of all new gyros, the transmitter, new memory, a new computer, the fine-guidance sensor, and new insulation. We deployed Hubble on Christmas Day. Hubble was back!

—JOHN M. GRUNSFELD

ABOVE: The crew of STS-103 poses for the traditional in-flight portrait on the flight deck of the space shuttle *Discovery*. In front, left to right, are astronauts Claude Nicollier, Scott J. Kelly and John M. Grunsfeld. Behind them, left to right, are astronauts Steven L. Smith, C. Michael Foale, Curtis L. Brown, Jr., and Jean-François Clervoy.

OPPOSITE: Astronauts C. Michael Foale (left) and Claude Nicollier are replacing one of the telescope's Fine Guidance Sensors. This is the second of three spacewalks on the STS-103 mission.

Astronaut Steven L. Smith retrieves a 35mm camera during the final spacewalk of the STS-103 mission. Many of the tools required to service Hubble are stored on the handrail attached to the robotic arm visible in the photograph.

Mission STS-109

The fourth Hubble servicing mission, STS-109, was launched on March 1, 2002, and lasted eleven days. In five spacewalks, the crew replaced the Faint Object Camera, Hubble's last remaining original instrument, with the Advanced Camera for Surveys (ACS), and performed upgrades and maintenance. The crew of space shuttle *Columbia* included Scott D. Altman, commander; Duane G. Carey, pilot; John M. Grunsfeld, payload commander; and Nancy J. Currie, Richard M. Linnehan, Michael J. Massimino, and James H. Newman, mission specialists.

STS-109 was my first spaceflight. I considered myself the luckiest space rookie since Alan Bean—Alan was the first rookie astronaut to walk on the moon and I was the first rookie astronaut given the opportunity to spacewalk on Hubble. What I remember most about that flight was how wonderful it was to leave the space shuttle in my big white spacesuit, work on Hubble, and view the Earth during my spacewalks. Our training was so good that I felt totally prepared to work on the telescope. I felt as if I had done my tasks a hundred times before. But nothing could have prepared me for the beauty of the environment in which I was working. While spacewalking I was able to see the entire Earth as a big ball. It was so beautiful I could hardly stand to look at it, more beauty than words can describe. I felt as if I was looking into heaven, that is how beautiful it was.
—MICHAEL J. MASSIMINO

ABOVE: Posing in front of the space shuttle *Columbia* upon their return is the STS-109 crew. From left are James H. Newman, Michael J. Massimino, Nancy J. Currie, Scott D. Altman, Duane G. Carey, John M. Grunsfeld, and Richard M. Linnehan.

OPPOSITE, TOP: Astronaut John M. Grunsfeld, anchored on the end of the space shuttle *Columbia*'s Remote Manipulator System robotic arm, moves toward the Hubble Space Telescope temporarily hosted in the orbiter's cargo bay on the fifth and final spacewalk of the mission.

OPPOSITE, BELOW: The horizon of a blue-and-white Earth and the blackness of space form the backdrop for this view of the cargo bay of the space shuttle *Columbia*, as seen through windows on the aft flight deck during the STS-109 mission. Visible in the bay are the gold replacement Hubble solar arrays.

ABOVE: With his feet secured on a platform connected to the Remote Manipulator System robotic arm, astronaut Michael J. Massimino hovers over the space shuttle *Columbia*'s cargo bay while working in tandem with astronaut James H. Newman during the STS-109 mission's second day of extravehicular activity.

OPPOSITE: Astronauts John M. Grunsfeld (right) and Richard M. Linnehan float in front of the Hubble Space Telescope at the close of the fifth and final spacewalk of the STS-109 mission.

Mission STS-125

The fifth Hubble servicing mission, STS-125, was launched on May 11, 2009, and lasted thirteen days. In five spacewalks, the crew installed the Cosmic Origins Spectrograph, replaced the Wide Field and Planetary Camera 2 with the Wide Field Camera 3, and performed upgrades and maintenance. The mission was intended to extend Hubble's operational life span until at least 2014. The crew of space shuttle *Atlantis* included Scott D. Altman, commander; Gregory C. Johnson, pilot; and Andrew J. Feustel, Michael T. Good, John M. Grunsfeld, Michael J. Massimino, and K. Megan McArthur, mission specialists.

My most indelible memory of STS-125 has to be that of watching the telescope disappear into the darkness of the sunset following deploy. Accomplishing this mission had been such a wild ride, from the mission cancelation after Columbia, later brought back to life after return to flight, assigned in 2006 with an April 2008 launch date, delayed to September, moved back up to August, delayed due to a hurricane, and finally put on hold for seven months due to a failure on Hubble. We had finished five straight days of spacewalks. Each spacewalk seemed to encounter a different obstacle, from bolts that would not turn until they almost broke to having to manually break off a handrail that had to be removed for the repair to succeed. Now we had just deployed Hubble with each task successfully accomplished. It had taken a team with people from all across the Agency working together both before and during the mission to bring us to this point. It was an incredible feeling, watching the telescope fade into black to continue its voyage of discovery, knowing the Hubble story would continue.
—SCOTT D. ALTMAN

ABOVE: The crew members for the STS-125 mission pose for a photo on the flight deck of the space shuttle *Atlantis*. In the front row, left to right, are Scott D. Altman, K. Megan McArthur, and Gregory C. Johnson. In the back row, left to right, are Michael T. Good, Michael J. Massimino, John M. Grunsfeld, and Andrew J. Feustel.

OPPOSITE, TOP: An STS-125 crew member onboard the space shuttle *Atlantis* snapped a still photo of the Hubble Space Telescope as it was grappled by the Remote Manipulator System robotic arm in preparation for servicing.

OPPOSITE, BOTTOM: The Hubble Space Telescope is pictured through an overhead window on the aft flight deck of the space shuttle *Atlantis* just prior to its release.

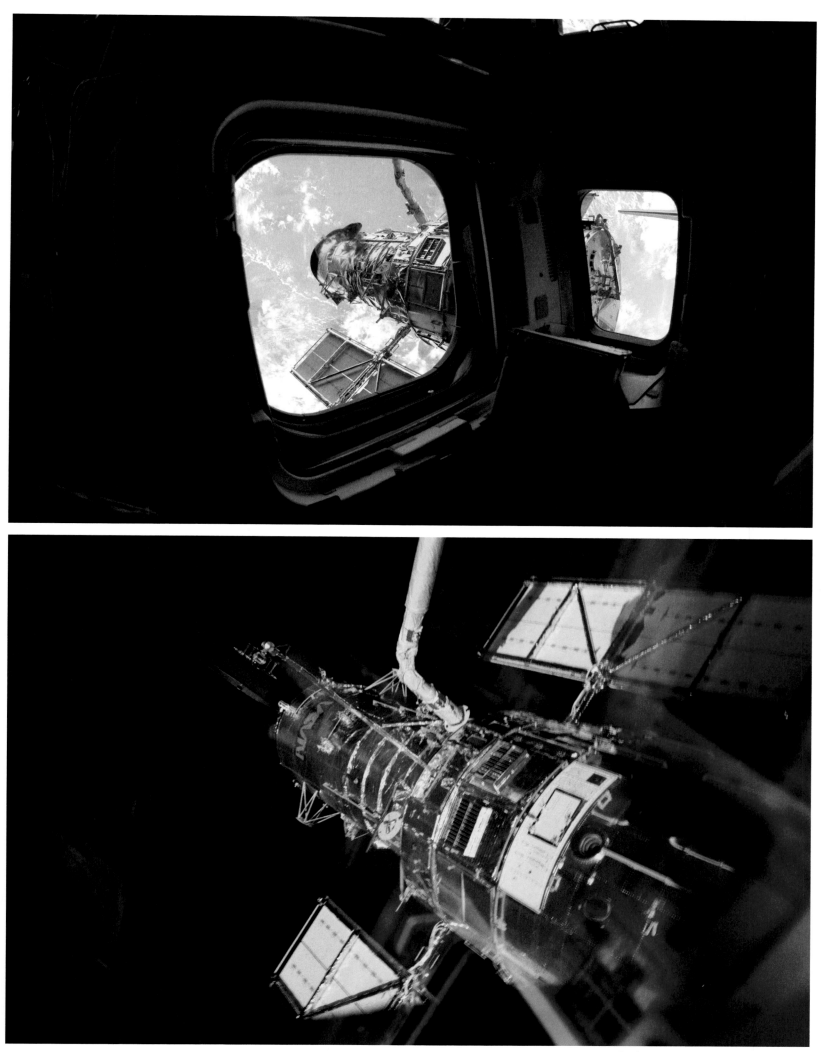

RIGHT: Astronaut John M. Grunsfeld, positioned on a foot restraint on the end of the Remote Manipulator System robotic arm, and astronaut Andrew J. Feustel (left) participate in the mission's fifth and final session of extravehicular activity.

OPPOSITE, TOP: Astronaut John M. Grunsfeld at work during the mission's final seven-hour-and-two-minute spacewalk.

OPPOSITE, BOTTOM: Astronauts John M. Grunsfeld (left) and Andrew J. Feustel work in the Earth's shadow on the mission's final spacewalk.

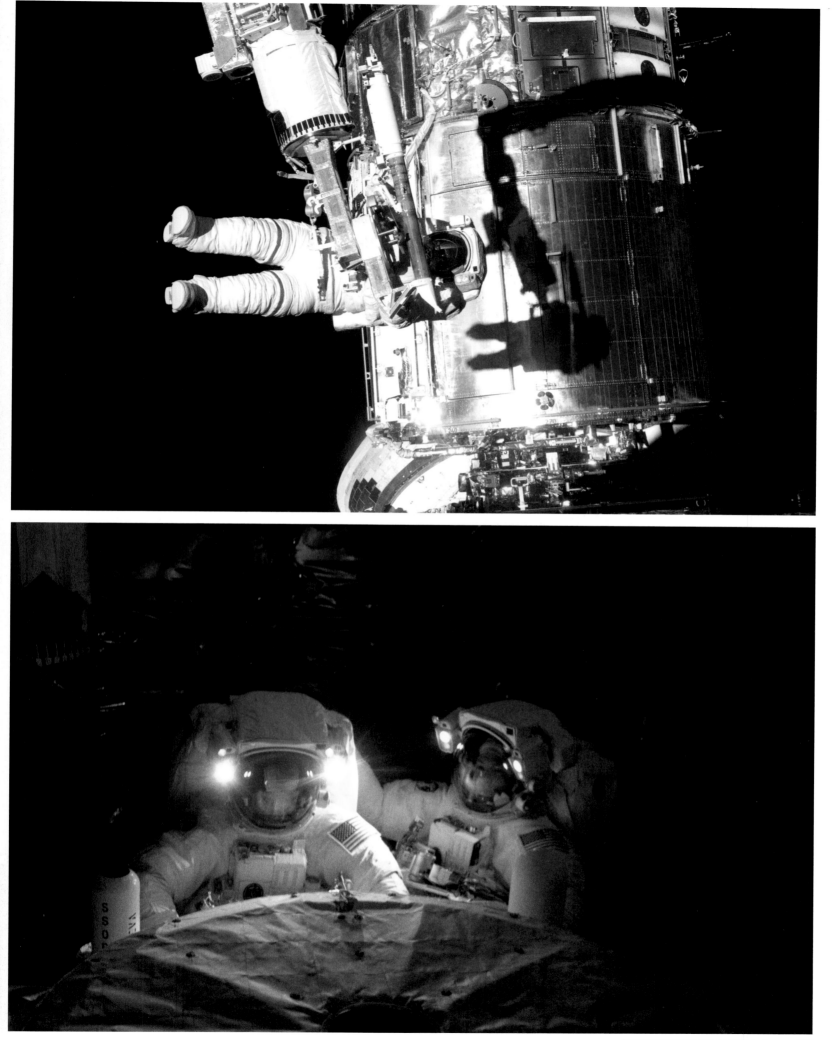

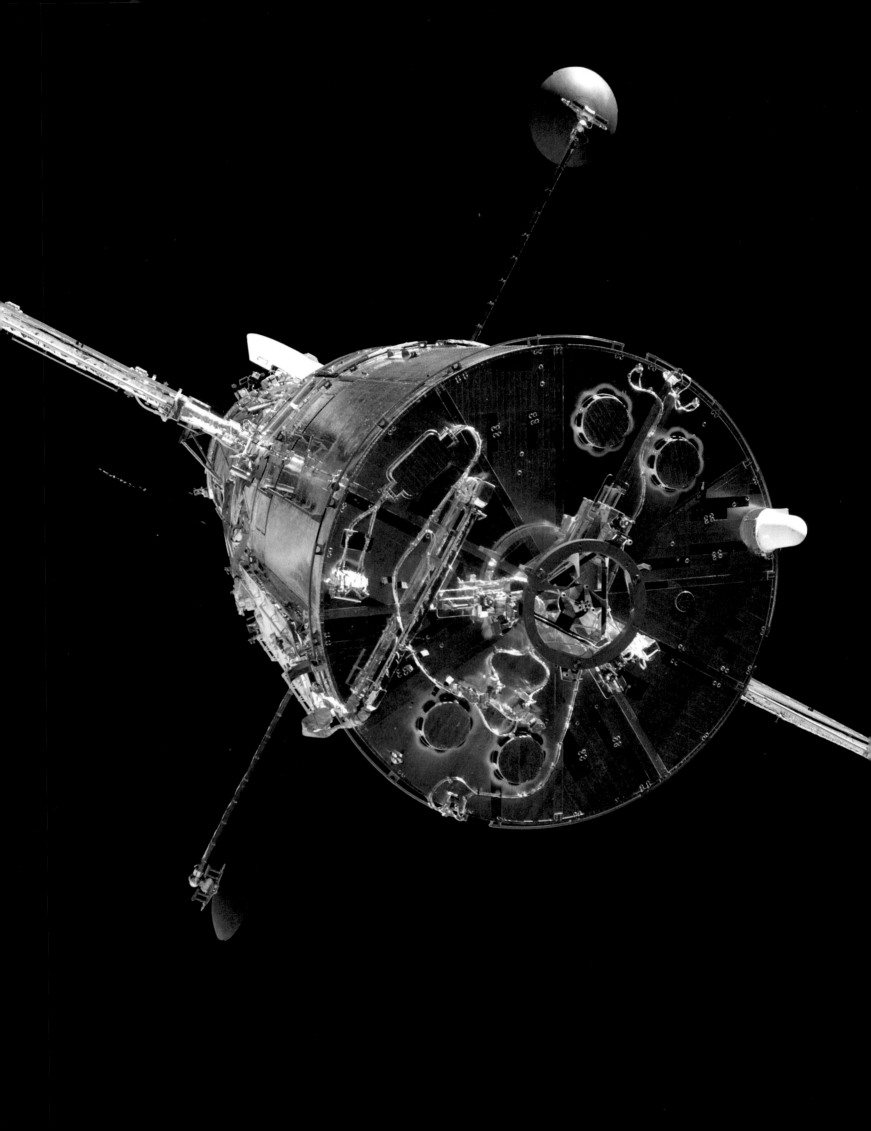

An STS-125 crew member aboard the space shuttle *Atlantis* captured this still image of the Hubble Space Telescope as the two spacecraft separated on May 19, 2009, after having been linked together for the better part of a week.

Epilogue
Edward J. Weiler

Since NASA was created in 1958, scientists have energetically planned to advance scientific knowledge and understanding of our solar system and universe. In 1978, astronomers were elated to have the first space observatory to be operated in real time. Called the International Ultraviolet Explorer, the observatory cataloged more than 104,000 objects, including comets, stars, interstellar gas, supernovae, galaxies, and quasars.

Of course, the best was yet to come for space astronomy. Astronomers now have far more capabilities to understand the wonders of the cosmos. The development of the space shuttle and other expendable launch vehicles enabled America to launch several state-of-the-art observatories, including the Hubble Space Telescope.

Hubble was the result of a herculean effort by thousands of dedicated individuals, including scientists, managers, engineers, support staff, NASA center personnel, contractors, international partners, and astronauts. While the last servicing mission in 2009 initiated a new beginning for the telescope, there is life after Hubble. There's a good chance Hubble will remain operational long enough to work in concert with its eventual successor, the James Webb Space Telescope, or JWST.

The JWST will be placed approximately one million miles from Earth, far beyond Hubble's low Earth orbit. In science terms, this is referred to as the Lagrangian point L2. Scheduled for launch in 2014, JWST's combined eighteen optical mirrors will be almost three times the diameter of Hubble's primary mirror. The larger diameter will collect more light to allow scientists to see farther and better than Hubble. In fact, scientists for the first time will be able to see back to almost the very beginning of time.

While the science community anxiously awaits JWST, more than a dozen spacecraft now in orbit are complementing Hubble observations. These include the Chandra X-ray Observatory and Spitzer and Fermi Gamma-ray space telescopes. Like Hubble, these missions are rewriting science textbooks. The Wilkinson Microwave Anisotropy Probe has made fundamental measurements of the properties of our universe as a whole. The Galaxy Evolution Explorer is generating an ultraviolet sky survey of galaxies. Kepler, launched in 2009, is the world's first mission to search for planets that orbit stars like our Sun.

Kepler's observations peer into what is called the "habitable zone"—the region around a star where the temperature is just right for liquid water. This is an essential ingredient for life.

The recently launched Wide-field Infrared Survey Explorer will allow scientists to survey the entire sky in the mid-infrared with far greater sensitivity than any previous mission. The spacecraft will image and catalog a vast number of astronomical objects. This will provide a unique storehouse of knowledge about the solar system, the Milky Way, and the universe. Next in the queue is the Nuclear Spectroscopic Telescope Array, or NuSTAR, currently scheduled for launch in 2011. The mission will expand our understanding of the origins and destinies of stars and galaxies and have five hundred times the sensitivity of previous instruments to detect black holes. It will perform cutting-edge science using advanced technologies and help to provide a balance between small and large missions in the NASA astrophysics portfolio. NASA also will be a substantial contributor to future international missions.

It is truly the best of times for astrophysics and astronomy.

What does the next decade hold for astronomy from space? That question will be answered by the National Academy of Sciences. It provides the Astronomy Decadal Survey, which, in essence, is NASA's astrophysics mission playbook. The survey will prioritize and provide guidance for NASA's decision process to plan and undertake future missions. The academy's assessment is based on scientific impact, technological readiness, and budgetary considerations as to what should be developed and launched first. Learning from missions such as Hubble, JWST, Kepler, and others, the playbook might propose projects to investigate the nature of dark energy, a repulsive force that is pushing the universe apart at an ever-faster rate. Other recommendations might point to the study of the physics of the Big Bang or missions to further investigate extrasolar planets.

The best and brightest engineers and scientists from NASA, industry, and academe have been tirelessly evaluating options for future missions. Certainly, there will be technical and budgetary challenges. I suspect NASA will continue to have multiple eyes in space, exploring the universe with more in development to expand our scientific frontiers for many decades.

Whatever happens, one lesson we have all learned from Hubble is that the number-one priority for future space observatories is to keep the science available to everyone. This is the American tradition of open competition to obtain the best. There also should be no limitations on releasing information and findings to everyone. The public has the right to see the data and, in turn, have it clearly explained.

As an astronomer, I feel very fortunate to be alive at this time when so much exciting research and exploration in space is happening. NASA's astronomy activities are integral to helping extend humanity's exploration activities. Astronomy will be at the forefront of extraordinary scientific advancement.

After twenty years of phenomenal discoveries from Hubble and other space observatories, our view of the universe and our place within it has changed forever. These observations have taught us there is still much to discover. I predict that we could very well, before the end of this century, prove definitively that there is life elsewhere. Indeed, as we push the frontiers of our knowledge and extend our exploration horizons throughout the solar system and beyond, the science of astronomy will be there to finally address that historic moment in answering the age-old question: Are we alone?

Clean-suited technicians, positioned on platforms, secure a solar array on the Hubble Space Telescope during assembly and testing procedures at the Lockheed Facility in Sunnyvale, California, in March 1988. Other technicians look on and one records the activity using a motion-picture camera.

Equipped with his five senses, man explores the universe around him and calls the adventure Science. —EDWIN P. HUBBLE

This book is dedicated to Hubble huggers everywhere!

ACKNOWLEDGMENTS

The editors want to acknowledge and express our deep appreciation for the cooperation and participation of the many scientists, engineers, astronauts, and mission managers who contributed to this publication.

We would especially like to thank Jon Morse, Astrophysics Division Director, NASA, Washington, D.C.; Michael Moore, Hubble Program Executive, NASA, Washington, D.C.; Eric Smith, Hubble Program Scientist, NASA, Washington, D.C.; Dan Woods, Director, NASA Science Mission Directorate's Strategic Integration and Management Division, Washington, D.C.; Ray Villard, News Chief, Space Telescope Science Institute, Baltimore, Md.; and Cheryl Gundy, Public Affairs Officer, Space Telescope Science Institute, Baltimore, Md.

Also, we would like to acknowledge the NASA leadership who gave us the time and opportunity to work on this project, including George T. Whitesides, Chief of Staff, Washington, D.C.; and David L. Noble, White House Liaison, Washington, D.C.

As always, we thank Abrams, especially Eric Himmel, Sarah Gifford, and Ankur Ghosh.

The Hubble Space Telescope is the result of thousands of dedicated workers in NASA, industry, and academe worldwide. In particular, we would like to thank the scientists, engineers, managers, and support staff at the NASA Marshall Space Flight Center; NASA Goddard Space Flight Center; Jet Propulsion Laboratory; European Space Agency; Lockheed-Martin Corp.; Ball Aerospace, Inc.; Goodrich Corp.; Space Telescope Science Institute; and the Association of Universities for Research in Astronomy, Inc. for their contributions toward making Hubble so successful.

Finally, we would like to give special thanks to all of the people who make up the National Aeronautics and Space Administration. Throughout history, we have gazed into the heavens with a childlike innocence and considered our place in the cosmos. Thanks to the people of NASA and their imagination, bold action, and, above all, perseverance, we have the Hubble Space Telescope to help us see and better understand our universe with renewed wonder and clarity.

PAGE 1: The Hubble Space Telescope has been returned to orbit after servicing in 1992. The crew of the space shuttle *Discovery* takes one last look at the satellite before returning to Earth at the end of the second Hubble servicing mission, STS-82.

PAGES 2–3: Astronomers encountered a mystery when they sought to determine the age of the open star cluster NGC 6791, one of the oldest in the Milky Way galaxy. Using Hubble to study the dimmest stars in the cluster, they uncovered three different age groups. Two of the populations are burned-out stars called white dwarfs: One group appeared to be 6 billion years old; another appeared to be 4 billion years old. The ages are out of sync with those of the cluster's normal stars, visible in this image, which are 8 billion years old.

PAGES 4–5: Looking westward, one of the space shuttle *Columbia*'s crew photographed the newly serviced and upgraded Hubble Space Telescope near the Earth's limb in 2002 at the end of the fourth Hubble servicing mission, STS-109.

EDITOR: Eric Himmel
DESIGNER: Sarah Gifford
PRODUCTION MANAGER: Ankur Ghosh

Cataloging-in-Publication Data has been applied for and may be obtained from the Library of Congress.

ISBN 978-0-8109-8997-9

Printed and bound in the United States of America
10 9 8 7 6 5 4 3 2

Abrams books are available at special discounts when purchased in quantity for premiums and promotions as well as fundraising or educational use. Special editions can also be created to specification. For details, contact specialmarkets@abramsbooks.com, or the address below.

115 West 18th Street
New York, NY 10011
www.abramsbooks.com